Gendered by Design? Information Technology and Office Systems

Gender and Society:
Feminist Perspectives on the Past and Present

Series Editor: June Purvis
School of Social and Historical Studies,
University of Portsmouth, Milldam,
Portsmouth PO1 3AS, UK

This major new series will consist of scholarly texts written in an accessible style which promote and advance feminist research, thinking and debate. The series will range across disciplines such as sociology, history, social policy and cultural studies. Before submitting proposals, a copy of the guidelines for contributors to *Gender and Society* should be obtained from June Purvis at the address above.

Out of the Margins: Women's Studies in the Nineties
Edited by Jane Aaron, *University College of Wales*, and Sylvia Walby, *London School of Economics*

Working Out: New Directions for Women's Studies
Edited by Hilary Hinds, *Fircroft College, Birmingham*; Ann Phoenix, *University of London*; and Jackie Stacey, *Lancaster University*

Making Connections: Women's Studies, Women's Movement, Women's Lives
Edited by Mary Kennedy, *Birkbeck College, London*; Cathy Lubelska, *University of Central Lancashire*; and Val Walsh, *Edge Hill College*, Lancashire

Mature Women Students: Separating or Connecting Family and Education
Rosalind Edwards, *South Bank University*

Gendered by Design? Information Technology and Office Systems
Edited by Eileen Green, *Sheffield Hallam University*, Jenny Owen, *UMIST, Manchester* and Den Pain, *Sheffield Hallam University*

Forthcoming

Feminist Politics and Education Reform
Madeleine Arnot, *University of Cambridge*

Women and Modernism
Gabrielle Griffin, *Nene College, Northampton*

Feminism, Sexuality and Struggle
Margaret Jackson

Subjects and Objects: Gender and Schooling
Jenny Shaw, *University of Sussex*

Friendly Relations: Mothers and their Daughters-in-Law
Pamela Cotterill, *Staffordshire University (Stoke-on-Trent)*

Aids: Gender and Society
Tamsin Wilton, *University of the West of England, Bristol*

Women in Britain 1914–1945
Edited by Sybil Oldfield, *University of Sussex*

Gendered by Design?
Information Technology and Office Systems

Edited by

Eileen Green
Jenny Owen
Den Pain

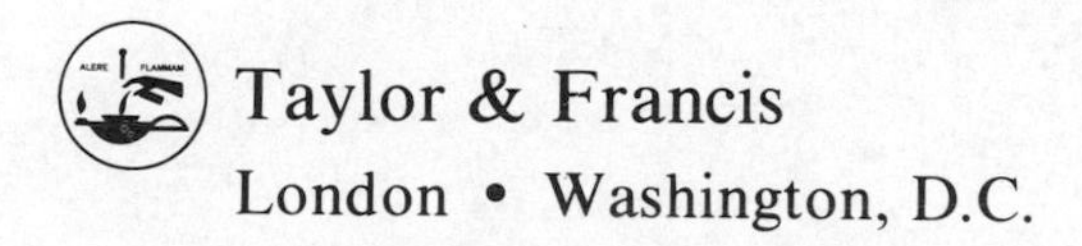

UK Taylor & Francis Ltd, 4 John St., London WC1N 2ET
USA Taylor & Francis Inc., 1900 Frost Road, Suite 101, Bristol, PA 19007

First published 1993

A Catalogue Record for this book is available from the British Library

ISBN 0 748400 915
ISBN 0 748400 923

Library of Congress Cataloging-in-Publication Data are available on request

Typeset in 9.5/11pt Times
by Graphicraft Typesetters Ltd., Hong Kong

Printed in Great Britain by Burgess Science Press, Basingstoke on paper which has a specified pH value on final paper manufacture of not less than 7.5 and is therefore 'acid free'.

Contents

Acknowledgments

The idea for this book emerged during one of our regular meetings of the Human-Centred Office Systems Project which was funded by the British Research Councils' Joint Committee on the Successful Management of Technological Change. Many of the ideas in the sections of the book written by ourselves were generated during the process of this research and we would particularly like to thank Dr David Budworth from the Joint Committee for his encouragement and support. We would also like to thank members of the project Advisory Group, several of whom are included as authors, for the stimulating discussions and feedback at project meetings, which sharpened our perspective. Special thanks go to Vivienne Mallinder who worked as a secretary to the project and spent many hours transcribing taped interviews and typing the manuscript. The production of the manuscript has not been an easy one, with patience and humour needing to be exercised by all those involved or affected by the process, including family and friends. Finally we are indebted to the many staff at 'City Libraries' who were both our collaborators in the research process itself and gave so generously of their time and ideas; and to the fieldwork carried out by Ian Franklin who worked as a research assistant on the project for three years.

Eileen Green, Jenny Owen and Den Pain
July 1992

Introduction

Eileen Green, Jenny Owen and Den Pain

> ... the men in our section were very sort of enthusiastic about it; whether it was good or bad, whether it worked or not, was *super*! ... suddenly you had to do *everything* on that piece of equipment and quite honestly sometimes we [women] used to sort of think: well how did we do the job before? Because they [the men] couldn't do anything unless they used that terminal ... (woman library assistant)

In this book we bring together two important and expanding areas of research and debate. First sociological analyses of gender relations in the workplace, most specifically in relation to information technology (IT); and second, the body of interdisciplinary research into the design of computer systems. The papers included in this international collection establish new links between empirical and theoretical work in computer science and social science, the most original aspect of which is the explicit development of gender perspectives on systems design processes.

Approximately one in three women in the UK labour force are employed in some form of 'office work' (EOC, 1991). Since the mid-1970s, office work has also been a major arena for the development and dissemination of IT. Early forecasts and debates often treated new office information technologies as vehicles of dramatic change. The optimistic versions suggested that the 'paperless office' would become a more 'modern', more satisfying and less stressful place to work; the more critical and pessimistic accounts predicted that many women office workers would lose their jobs, and that those remaining would face greater stress, increased pressure of work and potential deskilling. Interestingly, both versions implied a somewhat deterministic view of technology itself, in the face of which women office workers featured mainly as passive victims or beneficiaries (Webster, 1990, and this volume). Neither set of predictions can be said to have been fulfilled, and empirical studies reveal women office workers' own assessments of IT to be more mixed and more complex than either view originally envisaged (see for example, Liff, this volume).

Now that IT has ceased to be the 'new technology' of the 1970s and has become increasingly integrated into workplace environments and practices, the debates and perspectives have diversified. Many studies have indicated that the path of office IT application design and implementation is rarely a smooth one. In the mid-1970s, for instance, some studies found that up to 40 per cent of information systems could be described as failures (Hirschheim, 1985); studies published in 1991 painted a broadly similar picture (Hornby *et al.*, 1991; National Audit Office, 1991).

Within computer science, there have been diverging responses to these difficulties: an appeal to formal, mathematical methods of systems design, in some quarters, in pursuit of 'technical' control over complex software developments. In others, there has been the contrasting claim that it is the 'social' or 'user relations' aspects of the systems development process which have been marginalized by past design approaches, and which now need urgent attention (for a full discussion, see for example Friedman and Cornford, 1989, and Chapter 1, this volume). There is now increasing scope for interdisciplinary work on IT systems development methodologies, inviting collaboration between computer scientists, sociologists and anthropologists, as well as with the psychologists who have traditionally contributed to workstation and interface design (Murray and Woolgar, 1990).

Meanwhile in social science, too, the focus has begun to shift, towards a view of technology itself as 'socially shaped', rather than as producing good or bad impacts in a deterministic model. In this view, technology is not defined solely as 'object' (a car, or a computer), but also encompasses human knowledge and technique, and the forms of social relations within which these are used (Mackenzie and Wajcman, 1985). This 'social shaping' approach differs in emphasis from those which adopt a wholly 'social constructivist' position (Bijker, Hughes and Pinch, 1987; see also Murray and Woolgar, 1990). However, both invite a critical examination of the ways in which the boundary between 'social' and 'technical' processes or artifacts is negotiated, rather than accepting this as 'given' or taken for granted. From this angle, for example, the focus is on the *ways* in which the contrasting systems design approaches noted above have been defined: technical or mathematical (hard, rigorous, scientific), as against 'social' or user-oriented (soft, unscientific), and not simply on a comparison between the two.

The contributions to this volume all address the question of gender relations: a theme which has largely been absent from, or submerged within, evolving computer science and social science perspectives on the design of IT applications in the office context. Unsurprisingly however, the gender perspectives presented in the collection as a whole demonstrate differences of emphasis and theoretical direction.

Our own point of departure, for the research which prompted plans for this book, was an interest in extending the principles of 'Human-Centred Systems' research (HCS) to the women who make up the large majority of the typical office workforce. In Phase 2, the research team inherited a case-study organization[1] which offered exciting opportunities for innovative intervention in an emerging systems design process; we also (somewhat unconsciously) inherited a theoretical perspective with which initially several, and eventually all of us, were increasingly uncomfortable. The original perspective suggested a broad role for sociology within systems development, influenced by Braverman's work (1974) and the Scandinavian UTOPIA project (Bødker *et al.*, 1987). Issues of class based inequality were therefore more visible, which influenced the decision to focus upon definitions of skill and restricted training and career opportunities for women in IT-related areas. The gender element was (at this stage) more of an unworked out sub-theme, than the major perspective which it eventually became. Gender sits uneasily with the HCS approach which emerged partly in response to

Braverman's (1974) analysis of technology as tending to deskill and degrade labour in a capitalist context, although HCS has represented a radical strand in research concerning the design and use of computers. In the UK the HCS tradition has embraced both initiatives in creating 'socially useful production', for example through the 'Lucas Plan' of the 1970s and the London 'Technology Networks' of the 1980s (Cooley, 1980/1987; Collective Design, 1985); it has also included initiatives in devising computer systems to support and enhance human skill and discretion in manufacturing contexts (Rosenbrock, 1989; Gill, 1990).

Such initiatives have generated powerful critiques of existing uses of technology, as well as new prototype systems and products. In a sense, then, HCS sought to show that technology could indeed be 'socially shaped', to embody 'human' values broader than those of professional technologists, and opposed to those of the dominant multinational companies. But paradoxically, while challenging taken for granted views of technology, HCS tended to take 'human' as read. 'Human' was male, on the whole: typically a skilled craftsman in printing or engineering. Pam Linn (1987) has illustrated the difficulties experienced by a black woman dressmaker, in attempting to gain support from one of the London Technology Networks for a new project: the design, machining and maintenance skills involved were not recognized as 'technical'. 'Human' was also treated as unproblematic in a broader sense, as an implicitly homogeneous category, which left little scope to address issues of difference, conflict and identity, between or within workforces and communities, or to acknowledge the ways in which power may be experienced as enabling as well as coercive. Occupying a position 'on the margins', can offer opportunities to develop new insights, as a number of feminists have observed (see for example, Star, 1991).

The HCS tradition, then, was not one which could simply encompass the concerns and the experiences of women clerical workers in a new set of techniques or prototypes. As in other areas of social science, and of computer science, a redefinition of terms and boundaries was needed, rather than an attempt to assimilate women into a framework which had been defined largely in male terms (Hales, 1992).

Two current themes within gender analyses seem to us particularly important in this connection. First, there is now a substantial body of research documenting the ways in which gender relations are produced within the workplace itself, and especially in connection with technology (Cockburn, 1983, 1985). However, where office IT is concerned, the emphasis to date has been on the areas of training, use and implementation. We now need to investigate the ways in which gender relations may or may not be important, and perhaps be susceptible to intervention and change, during the office systems design process itself.

Second, recent feminist research on information technology has begun to prise apart some familiar associations. Donna Haraway challenges us to accept that the boundary between 'human' and 'non-human' is neither secure nor absolute; to this end, she evokes the 'cyborg' concept, part organism, part machine, not as a science fiction monster, but as a positive image of creative interdependence (Haraway, 1991). Or as Cynthia Cockburn has put it:

> We do not need to have a heart pacemaker or be plugged into a kidney machine to be a cyborg. If you took away my glasses, my pen, my telephone, wordprocessor, cooker and car, let alone the medical, nutritive,

> transport and communications systems to which these connect me, I would not be identifiable as the person I am. The subjectivity of everyone in industrialized societies, man or woman, engineer or childminder, is a technological subjectivity. (1990, p. 11)

These emerging perspectives encourage us to let go of long established oppositions between 'human' (good, flexible, fallible?) and 'technology' (threat or promise? but anyway 'outside' us). They invite us to develop new analyses, and new strategies through which, in Cockburn's phrase: 'to uncouple technology and domination'.

A major part of such 'new' analyses comes in the form of the gender perspectives outlined in the following chapters, many of which demonstrate a potential for developing opportunities for active intervention in the office systems design process, opportunities which could empower women workers and make fully visible to both themselves and their organizations, the nature and range of skills and competencies which arise from traditionally undervalued 'women's work'. The first section sets the scene: reviewing existing interdisciplinary approaches to office systems design, and exploring the broad issues raised by developing gender perspectives on IT. Section II moves on to highlight both some of the major theoretical issues which underlie discussions of gender and computer systems development, and two specific areas in which significant debate is under way.

The third section presents a contemporary view of IT in the office context. This includes both a broad view of changing trends and working patterns, and a close look at the ways in which the most common forms of office related IT equipment reflect gendered assumptions, in both their design and use. The final section presents accounts of new initiatives in connection with the design of office systems. This is the longest and most detailed section which incorporates an international perspective. The four case-studies presented illustrate a range of new approaches currently being developed: two in Britain, one in Scandinavia and the other in Italy. The aim here is to supplement the overviews set out in the first two sections with detailed case-study material. This material facilitates in-depth analyses of the successes, the failures and the broader implications of office systems design approaches which aim to make gender perspectives an integral component.

The first two chapters attempt to provide the reader with a broad overview of the two major areas addressed in more detail in the following sections. In Chapter 1 we concentrate upon an examination of the nature of human-centred design, in comparison with other major approaches to the design of office-based computer systems. A set of design principles are proposed, raising a range of issues which are echoed and expanded upon in subsequent chapters. The most notable of these are: male tenure of technical knowledge, which is analyzed in Section II; the expansion of areas of work now amenable to technological interpretation, followed up in Chapter 8; and finally the question of the most appropriate methods and techniques to adopt in the creation of human-centred approaches, a theme explored in different ways by many of the authors. More generally, Chapter 1 injects a note of caution in recognizing the *limits* of participatory design, limits largely of an externally imposed nature which threaten to marginalize innovative ideas and practices: a prudent caution which is underlined by events outlined in several of the case-study chapters, but happily fails to dampen the bubbling enthusiasm for collaborative efforts at change!

In Chapter 2, Flis Henwood is concerned with the relationship between theory and change. The chapter sets out to analyze the relationship between the concepts and frameworks with which we seek to understand the gender and IT relationship, and the strategies we develop to enable us to transform the presently gendered relations of technology. Via a broad review of the literature, two major approaches to the development of gender perspectives on IT are identified. Critical of these approaches, she advocates a wariness of the 'add women and stir' approach, arguing instead for frameworks which address the diverse ways in which technology and gender interact.

The chapters in Section II attempt to do just that. Chapter 3 by Susanne Bødker and Joan Greenbaum, suggests that office-system design approaches should concentrate upon people rather than upon 'things', and preferably gendered people. Using research from a gender perspective, the authors explain why and how cooperative design can be used to enable systems developers and office workers to work together when designing applications which will support working practices. Examples from the authors' own research and experiences in Scandinavia and the United States are provided, to demonstrate that there is realistic hope for new and effective approaches. In Chapter 4 we turn our attention to what Fergus Murray introduces as 'a handful of jigsaw puzzle pieces' towards an analysis of the relationship between masculinity and technology. Drawing upon two major pieces of research, the author explores male tenure of technology, and the ways in which aspects of technology define masculinity. Having analyzed the relationship between the 'project' type style of systems development, which thrives upon the 'Boy's Own' heroism component of masculine cultures, three broad perspectives are proposed. Categorizing these as the cultural, the structural and the psychoanalytical, the author concludes with an attempt at theoretically 'unpicking' the mutual interdependence between masculinity, science and technology.

Chapter 5 focuses upon a particular area of the computing field, expert systems, exploring the broad issues surrounding expert systems and gender. Alison Adam and Margaret Bruce follow a brief history of Artificial Intelligence (AI) and an explanation of expert systems, with an examination of some of the more philosophical questions related to the type of knowledge or expertise represented in expert systems . Having investigated the background to the language used in this area of computing, and analyzed its appeal for women thinking of computing as a career, the authors draw upon their own research in representing women's views on expert systems. The chapter concludes by noting that 'the introduction of expert systems cannot be regarded as neutral with respect to gender', but *could be* designed to make more opportunities available for women.

The two chapters in Section III address the major theoretical perspectives on technological change and office work. Sonia Liff in Chapter 6, combines analyses of technological change and gender relations in the office to provide a critical examination of the above. Using material from a range of existing studies, and detailed findings from her own research on women workers in the West Midlands of Great Britain, the author argues that women office workers' experience of newly implemented IT is both complex, and at times, contradictory. The chapter explores the ways in which current technological change in the office has affected the content of women's work and also the experience of doing such jobs, without significantly disrupting the boundaries of gendered occupations. Her approach concentrates upon two aspects of office systems design: the scope of the technology introduced and the ways in which systems are being implemented, concluding that in most cases the women users are the last to be consulted, if they are consulted at all. There is a need for employers to rethink job requirements and restructure jobs across existing occupational boundaries, but radical initiatives of this kind are more likely to come from groups with an explicit commitment to equal opportunities than traditional managers.

'Women's Work' is also the main theme of Chapter 7. Here Juliet Webster turns to a detailed analysis of the word processor, a key component of office automation. Having considered the trajectory of the development of word processing technology, showing how the design and evolution of the equipment has diverged from its original path, the author demonstrates that social structures and relations within workplaces may confound the direction of technological change. The core of the chapter is an examination of the impact of word processing systems upon the area of office work most fundamentally affected: women's secretarial and typing work. Supplementing Sonia Liff's broad-based analyses of technological change and office work, Juliet Webster explores the original forecasts of the impact of word processing upon work organization and skills, comparing them with the actual experience of automated word processing a decade later. The chapter concludes with the suggestion (backed by the author's own research data) that simple unilinear accounts of the development of technological and organizational change are inadequate.

The final section on new initiatives in Europe and Scandinavia opens with our own Chapter 8, which is based on an account of an 'action research' project on human-centred systems design. We begin by briefly reflecting upon some of the tensions (both creative and constraining), which proved to be a feature of both the research process itself and the case-study organization. The latter and the major stages of the research process are described in detail in the second part of the chapter, after which we move on to outline the two major approaches to systems design adopted: a study-circle method (initiated in Scandinavia) and a long term, broad based design team. The next section evaluates the strengths and weaknesses of the design initiatives, including the unresolved tensions exposed by the research, and is followed by a detailed analysis of gender issues in relation to user involvement. Finally, we propose a series of guide-lines in relation to human-centred system design strategies, guide-lines which incorporate a gender perspective.

In Chapter 9, Mike Hales and Peter O'Hara also reflect upon the strengths and weaknesses of 'participation', in a systems design project in the sphere of local government. Stressing that the chapter draws upon 'learning in a live setting', and that the methodological framework includes methods and interpretations 'invented as they went along', the authors tell us in the first section why a participatory approach was supported. Like Chapter 8, it is presented in a deliberately 'unreconstructed' form, which demonstrates the authors' awareness of the fact that it is often the 'raw' details of the research process which are the most informative for others engaged in parallel enterprises. Study circles (among others) were used as a method for empowering the design participants, together with prototyping as the main developmental technique. The successes and failures of the approaches are then described, with the authors recognizing the difficulties posed by the 'bottom-up' nature of participatory design within a traditional, top-down, bureaucratic organization. The final section deals with knotty issues in design politics posing a number of questions in the areas of empowerment and control, academic research and 'learning by doing', and the practice of equal opportunities. Chapter 10 continues the 'learning by doing' theme, drawing upon the extensive experience of teaching and action research programmes gained in Italy by Fiorella de Cindio and Carla Simone. In this chapter the authors propose the 'Universes of Discourse' as a tool for analyzing the various dimensions of work and its computerization. The first part explains and illustrates in detail the Universes of Discourse, while the second part describes their application in training and in the analysis and (re)design of the organizational settings involved in affirmative action programmes. This section also contains a detailed description and evaluation of the framework in use in teaching and action research programmes devoted to women. The chapter concludes with the authors' reflections on the particular insights gained from working with women-only groups. They argue that adopting women's work and viewpoints as the entry perspective for analysis and (re)design of organization and computerization reveals 'the organization as it *really* is'.

The final chapter in this last section is a 'project in progress' paper, which draws upon an action research project carried out by Eva Avner in Sweden, in collaboration with a Swedish clerical workers' union. The chapter opens with a brief overview of trends in Swedish women's office employment, and trade union policies on computerization. In the second part, an example of an innovative trade union project designed to support and improve opportunities for women office workers is discussed in detail. The chapter echoes many of the themes on women's clerical work presented in preceding chapters, concluding with the equally familiar view that employers and others frequently underestimate the skills and radical attitudes of women office workers.

The reader will discover that themes such as those introduced above represent a key focus in many of the papers contained in this collection; however all of the authors included make gender relations a focal point of their analysis, rather than an afterthought. We thus anticipate that this volume will make a significant contribution to a field which has evolved into an exciting, challenging area of interdisciplinary work.

Note

1 The Human Centred Office System Project, based at Sheffield City Polytechnic, spanned and was funded in two phases of development, by the UK ESRC/SERC Research Council's Joint Committee on the Successful Management of Technological Change. Phase 1 (1984–87) and Phase 2 (1987–91) incorporated different research teams, although the case-study organization remained the same. Only two members of the original team worked on Phase 2 until the end of the project.

References

Bijker, W.T., Hughes, T. and Pinch, T. (Eds) (1987) *The Social Construction of Technological Systems*, Cambridge, MA, MIT Press.

BØDKER, S., EHN, P., KAMMERSGAARD, J., KYNG, M. and SUNBLAD, Y. (1987) 'A Utopian Experience — On design of powerful computer-based tools for skilled graphic workers' in BJERKNES, G., EHN, P. and KYNG, M. (Eds) *Computers and Democracy — A Scandinavian Challenge*, Aldershot, Avebury.

BRAVERMAN, H. (1974) 'Labour and Monopoly Capital', New York, Monthly Review.

COCKBURN, C. (1983) *Brothers: Male Dominance and Technological Change*, London, Pluto Press.

COCKBURN, C. (1985) *Machinery of Dominance: Women, Men and Technical Know-How*, London, Pluto.

COCKBURN, C. (1990) *Technical Competence, Gender Identity and Women's Autonomy*, paper prepared for the World Congress of Sociology, Madrid, 9–13 July.

COLLECTIVE DESIGN/PROJECTS (Eds) (1985) *Very Nice Work If You Can Get It — The Socially Useful Production Debate*, Nottingham, Spokesman.

COOLEY, M. (1980) *Architect or Bee?* Slough, Langley Technical Services. (Revised version published in 1987).

EQUAL OPPORTUNITIES COMMISSION (EOC) (1991) *Women and Men in Britain 1991*, London, HMSO.

ERIKSSON, I.V., KITCHENHAM, B.A. and TIJDENS, K. (Eds) (1991) *Women, Work and Computerization: Understanding and Overcoming Bias in Work and Education*, Amsterdam, Holland.

FRIEDMAN, A. and CORNFORD, D. (1989) *Computer Systems Development: History, Organization and Implementation*, London, Wiley.

GILL, K.S. (1990) *Summary of Human-Centred Systems Research in Europe*, report from Seake Centre, Brighton Polytechnic.

HALES, M. (1992) 'What has Gender to Do with Human Centred Design? or Why Don't We Do More Things Different?' (draft) paper for the CIRCIT Seminar Series, Gender, Work and Technology, Melbourne, Australia, August.

HARAWAY, D. (1991) *Simians, Cyborgs and Women — The Reinvention of Nature*, London, Free Association Books.

HIRSCHHEIM, R. (1985) *Office Automation*, Chichester, Wiley.

HORNBY, P., CLEGG, C.W., ROBSON, J.I., MACLAREN, C.R.R., RICHARDSON, S.C.S. and O'BRIEN, P. (1991) *Human and Organisational Issues in Information Systems Development*, Paper from the MRC/ESRC Social and Applied Psychology Unit, University of Sheffield.

LINN, P. (1987) 'Gender Stereotypes, Technology Stereotypes' in MCNEIL, M. (Ed.) *Gender and Expertise*, London, Free Association Books.

MACKENZIE, D. and WAJCMAN, J. (Eds) (1985) *The Social Shaping of Technology*, Milton Keynes, Open University Press.

MURRAY, F. and WOOLGAR, S. (1990) 'Social Perspectives on Software', Unpublished report produced for the ESRC '*Programme on Information and Communication Technologies*' (PICT).

NATIONAL AUDIT OFFICE (1991) *Information Technology in the National Health Service*, London, HMSO.

ROSENBROCK, H. (1989) *Designing Human-Centred Technology: A Cross-disciplinary Project in Computer-aided Manufacture*, Amsterdam, Springer-Verlag.

STAR, S.L. (1991) 'Invisible Work and Silenced Dialogues in Knowledge Representation' in ERIKSSON, KITCHENHAM and TIJDENS (Eds) pp. 81–92.

WEBSTER, J. (1990) *Office Automation*, London, Harvester.

Section I

Context and Overviews

Chapter 1

Human-Centred Systems Design: A Review of Trends within the Broader Systems Development Context

Den Pain, Jenny Owen, Ian Franklin and Eileen Green

Introduction

In this chapter we are interested in locating human-centred design within the broad spectrum of approaches to the development of office computer systems. We start by taking a brief look at systems development in general, considering how things have changed over recent years. Our analysis is continued by reviewing the major approaches to systems development and noting their relevance to human-centred design. We then concentrate on the significant milestones within the human-centred field before expounding our own particular view. We conclude by predicting some of the future trends for human-centred design.

Background to Systems Development

Before we can look at how the methods used to develop computer based information systems have changed in the thirty or so years from the late 1950s to the 1990s, we need a brief analysis of how the systems themselves have changed. Our particular area of interest is that of commercial computing (office systems), or data-processing as it was called in the 1960s; however, most of what we have to say also applies to other areas of computing, such as military applications and production line computing in the manufacturing sector.

Although much has changed, it is important to recognize that there are many fundamental ideas that are as salient today as they were in the 1950s. This stability challenges the spurious arguments regularly advanced by computer specialists which suggest that new technologies or new methods can solve many difficulties and problems by virtue of their very newness. Computer systems are still basically concerned with the automation of tasks which have in the past and still could be, performed by humans. What *has* changed since the late 1950s, is the scale and extent of this automation. Computers now appear in most areas of working life from the smallest one-person firm, to the largest multi-national corporation, whereas in the early days of computing they were only to be found in a few large corporations, banks, insurance companies, government offices etc. and then only doing a few applications — payroll, ledger-accounting and stock control.

The nature of information systems has remained largely unchanged; they are composed of software, hardware, data and the associated manual procedures and organizational context. Information systems continue to make use of the fundamental abilities of computers, i.e. the capacity to store vast quantities of data, the provision of speedy access to this data and the ability to process it (perform comparisons, do calculations, etc.) extremely rapidly. Currently, information systems are much more complex, with the integration of many functions within one system, and the ability of data communications to link many machines together over a wide area. It is this increased complexity, together with their widespread proliferation, that gives information systems such a significant role in the quality of most peoples' working lives.

Some examples may help to illustrate these points. Early commercial systems were often small scale (sometimes single programs) in areas classically suited to computing, for instance payroll where the same procedure and calculations needed to be carried out many times on similar pieces of data. Today's systems can include large scale (hundreds of programs and machines) integrated systems such as airline reservations, where customers all over the world can browse possibilities from the same database simultaneously with airline companies trading with each other, buying and selling flights and services.

The development of information systems, whether large or small, remains the province of humans and continues to be regarded as a specialist task, a perception which has persisted since the early days of commercial computing. It is the relationship between these specialist developers and others concerned with the emerging information system, which is at the heart of our deliberations in this book. Obviously much has changed in the way information systems have been developed since the late 1950s. Friedman and Cornford (1989) provide an excellent historical account of systems development in Britain. Their analysis suggests three phases, where the paramount factor restraining further computerization has changed from hardware constraints through software constraints to one of problems with user relations. Our own research confirms the notion of this third phase. For example, in our case-study of a particular local government systems development process, we observed that the first act of the steering committee was to appoint a design team. This team included (and was headed by) professionals from the user department as well as analysts and designers from the central computing services section. (A full account of this case-study research is provided in Owen, 1992, and in Green, Owen and Pain forthcoming).

At a more general level, the concern in commercial computing with relations between users and specialist developers can readily be seen when we briefly trace the kinds of development methods in use from the 1950s to the 1990s. The first (1950s) methods of systems development consisted of little more than programming, accompanied by limited discussions with users about the inputs and outputs, and the necessary calculations. There was, of course, very restricted user-choice in that input was via punched cards, data was stored (if at all) on magnetic tape and output was printed on paper. As user expectations increased and the technology developed, information systems obviously became more complex, consequently including the process of ascertaining users' requirements, designing data-structures and more difficult screen layouts, etc. Hence the tasks of systems analysis and design became part of the development process along with programming.

With the continued increase in both the scale and complexity of information systems, came the need to control and manage the actual process of systems development and the (now considerable) numbers of people involved. This led the major organizations involved with systems development, education and training, to propose standard (and often structured) ways of doing systems development, which are usually referred to as systems analysis and design methods or methodologies. These standard methods date back to the 1960s (e.g. RAND Corporation USA) and have evolved to form the plethora of methods currently regarded as 'standard'. For example, within the EC some of the most common ones by country are: Britain SSADM, France MERISE, Netherlands SDM, Italy Dafne, Germany Vorgehensmodell (Computing, 1991). The use of these standard methodologies has not, however, prevented the development of many inefficient, poor quality information systems. Therefore many alternative approaches to systems development have been created over the years; most are directly concerned with the relationship between the specialist developer and the users of the information systems. Two fundamental questions come to mind when analyzing this 'user-relations' phase: who are the users, and what is the nature of their relationship with the analysts and designers? In the next section we will examine the major trends within systems development.

Current Trends in Computer Systems Development

The Environment of Traditional and Formalized Methods (within the UK)

By and large, the main responsibility for systems design rests with those labelled as 'experts' in technical or computing knowledge. That is, people who have been trained in such skills as database design, program design and systems design, who will probably also have experience of fact-gathering, interviewing and so on, even if they have no actual formal training in these areas. This group of 'experts' are predominantly white, middle-class men, usually with an educational background in science. Those who have formal computing qualifications will have encountered the social sciences, but will almost invariably have been encouraged to use them as a way of analyzing the *consequences* of technological design, rather than as a series of perspectives to analyze the design process itself. There are numerous examples of social science being used to explain why people at work have become alienated from and frustrated by particular designs. The seemingly inevitable conclusion is not that there is anything radically wrong with the methods or approaches to design but that the individual users (or groups) need to be trained in the existing methods before they can fully appreciate the 'brilliance' of the designs. There are particular situations where it appears, on the surface, that workers other than the technologists have control over the development, such as when user-managers lead the design team. However the surface nature of this appearance is usually revealed, demonstrating that it is the computer specialists who have the real control. In one particular systems development process we witnessed the technologists refer to the users in a manner which suggested that they could be manipulated as puppets by the technologists, even though in theory, the users were in charge.

> We're not exactly going to ask the users, are we, it's more saying to them 'this is what's going to happen!' All right, it might be a bit two-way . . . (D. Smith, Computer Services Manager; quoted in Owen, 1992)

This notion of where the 'real' power lies is borne out by noticing that it is the computer scientist's methods and language which are most often used. In the case-study referred to above, the user-manager, in a courageous bid to maintain control learnt more about a particular design method in a couple of months than the analysts with all their training courses did in a year.

Many organizations recognize that the development of computerized information systems has important consequences for career development, health and safety, office layout and ergonomics. Consequences will be highlighted by clerical trade unions who recognize the significance of the impacts of such developments on their members' working lives. Even in organizations which do recognize the significance of the changes brought about by the introduction of automation into office work, the organizational aspects are often underestimated. The concerns of occupational psychologists and ergonomists over such aspects as the quality of working life and job satisfaction, are generally regarded as peripheral to the main systems design effort, which is dominated by technical interests centring around the efficiency of the system. Examples of this can be seen where such elements as human factors (discussed later) are bolted on to the existing computing methods of systems design (Damodaran *et al.*, 1987).

It is our view that when it comes to developing information systems in the area of clerical work it is the computer specialist who has the real power and control over the design process. This matters because it is clear to us that with the continued introduction of automation into office work (especially with the advent of wide-ranging 'integrated' information systems) the influence and consequences of such computerized systems is significantly changing all areas of people's working lives. This is particularly important when it affects what is defined as 'technical' and what is defined as 'social' or 'organizational', because the more debates that are defined as technical, the stronger the influence technological experts (a relatively small group of computer scientists) will have over most people's lives in relation to both the nature and quality of paid work.

A good example of the shifting of the focus of discussion from the organizational arena to a technical one occurred in our case-study (see Chapter 8, this volume). A debate arose in the design team as to whether a communication from the organization to its clients should be dealt with on a local branch basis or sent out from the central headquarters. Before automation, this debate would have been regarded as one about organizational policy and the concerns of the female clerical workers (at the branches) to safeguard aspects of client care, would have been held in high regard. What actually took place (in the context of the development of a new information system) was a debate between the technologists, who argued for a centralized solution on the basis of better design and use of expensive equipment, and the women clerical workers who found it difficult to present their preference for a distributed solution in equally powerful terms. Now, it is not just that the male technologists were able to enter the debate on the back of a technological issue, but also that their 'scientific' arguments were held to be as significant as those arguing from the basis of organizational policy. This example serves to show how increasing areas of work are being defined as open to technical interpretation, and that when that happens the arguments of the technologists — developed as they are in very different fields — are often seen as the most cogent.

Computer Science, Engineering and Professionalism

Let us return to the group of technologists and examine the nature of their experience. As has already been pointed out, this group is predominantly male with an educational background in science. Professionally this group allies themselves with engineers, an alliance which is evident in the increasing use of the terms 'software engineer' and 'systems engineer', rather than the former terms of programmer and analyst. The engineering title is also evident in moves by engineering professional bodies such as the IEE (Institution of Electrical Engineers) to ratify computing courses at universities in order to try and establish computing specialists as professionals. There are also very obvious parallels with other forms of engineering. In practice the computing scientist uses a basis of some theory plus a strong pragmatic 'if it works it must be OK' element. All of this could be seen as linked with 'professional protectionism' or occupational closure (Johnson, 1972), that is, only those suitably qualified and experienced shall be allowed to do this job. This practice might be acceptable if the job was simply the introduction of automation into an area dominated by white male professionals, but as we have frequently observed, automation is brought into areas of clerical work where women predominate. However, such 'unconscious' patriarchal practices which privilege traditional 'expert' forms of knowledge, also serve to reproduce gendered patterns of clerical work, albeit in a new form. We are not suggesting that such reproduction of gender divisions is consciously designed in, but rather that the systems engineer, with his scientifically based tools, tends not to consider workplace relations and career structures to be elements of the design. The 'natural consequence' of his designs is therefore a reproduction of gender divisions from the engineering domain to the clerical.

Some insight into this can be gained through an analysis of current systems design methods, for example, the use of a group of popular structured approaches which utilize concepts such as information on a 'need to know' basis, deliberately restricting the abilities of some jobs (and people) to see or understand their place within the whole information system. What better method for producing dull routinized jobs with no possible career paths?!

Our own previous research has illustrated how difficult it can be for women to break these technological gender barriers (Owen, 1992). In a large engineering firm, for instance, we found that some women clerical workers had played an active part, informally, in information systems development processes, thus gaining substantial IT expertise. However, as their 'function' was still defined as clerical, they were not able to gain recognition for this expertise, nor to find opportunities at work to extend the use of these newly-found IT skills. As one of the women concerned commented:

> It's unlikely to change. I've made a rod for my own back, in one way. Your movement's not that free round the company; he [the departmental manager] doesn't want to lose someone who's making the system work. In some ways, they force women out of the company; women get as though they don't want to try. I wouldn't have the same enthusiasm after a few more years. (Louise, clerical worker)

The question our argument leads us to, is: is it possible for human-centred methods of design to open up the design process? That is, rather than reproducing existing class and gender divisions in the workplace, can human-centred approaches challenge taken for granted assumptions about the ways in which the 'technological' and 'social' or organizational domains are understood, during processes of systems development? We attempt to address this question more fully towards the end of this chapter and in Chapter 8.

Formal Methodologies

Formal methodologies are the most widely applied of the computer science methods. These are the so-called 'hard scientific' ways of doing computing characterised by a range of methods such as IBM's VDM (Vienna development method), the British government's SSADM (structured systems analysis and design method) and object oriented analysis and design (Coad and Yourdon, 1991a, 1991b). Structured methodologies for analysis and design such as SSADM (Cutts, 1987) and that of DeMarco (1978), attempt through several stages to produce a logical solution to a system's design. For Yourdon (1986), for example, system design is seen to be a rational decision-making process carried out on behalf of management.

Despite some long held prejudices by technocrats, the context within which computer systems are used is still a human and organizational one, even some of the most technically based methods have acknowledged this. In his book *Object-oriented Software Construction* Meyer (1988) states,

> what matters is the external factors (ones perceptible to users and clients), but they can only be achieved through the internal factors (perceptible to designers and implementors).

The rest of the book is of course taken up with an explanation of these internal facets of software design.

A recent movement can be seen in the development of even more formalized and mathematical methods of system design. Mathematically based formal specification methods such as VDM and Z have been developed with the admirable aim of producing flexible and reliable software, but their emphasis on provability and mathematics makes them a tool of the programmer/designer, not of the user, unless of course, we are to expect users to add the intricacies of higher mathematics to their existing expertise.

All of these formalized ways of developing computer systems have several attributes in common: all assume a rational scientific paradigm for their area of work; all have emerged from a very male dominated culture and work experience;

and all have been developed from the software engineering (programming) side of the computing spectrum. These 'structured' methodologies do emphasize that communication with the user is important, but the various techniques (data-flow diagrams, entity-relationship models etc.) are obviously geared up to making the production of programs and their associated database a simple and straightforward process. These techniques for 'communication' do not fit well with users' existing skills and experiences, making their participation unequal to the 'expert' system designers.

Prototyping

As a method for systems development, prototyping often comes under the heading of an experimental technique. The concept is a familiar one as it has been used in many fields for a long time, for example, the architect's scale model, the mock-up car for tests in the wind tunnel, or the trial run of a new pudding before a dinner party. It can be defined in the following manner: a model (prototype) of the information system is built, evaluated and refined until all parties are satisfied that the proposed information system (as represented by the model), will meet the requirements within the agreed constraints.

From the early 1980s, prototyping has received increasing interest from many fields of computing and there is an extensive literature on the subject — see for example Eason (1987), Dearnley and Mayhew (1983) and Floyd (1984). This form of development has resulted both from the computer industry's desire to increase the productivity of programmers, and from parallel dramatic improvements in hardware performance and cost reductions characteristic of the late 70s and early 80s. With the introduction of 4GLS (fourth generation languages) and more powerful computers it is now possible for prototypes to be developed quickly and cheaply enough for evaluation and refinement to be feasible. Before 1980 this was rarely the case, and it was a standing joke (not enjoyed by programmers and designers) that the computer industry was the one place where people paid full price for, and had installed, the prototype. Prototyping by its very nature requires significant active response from the user community, and has therefore been seized upon as 'the answer' by theorists interested in the 'user end' of the development spectrum. It seems somewhat ironic however, that a tool developed by software engineers to improve the productivity of programmers is now seen by many as the way for users to wrest control from the technologists.

Indeed some proponents of 4GLS have gone so far as to suggest that this development marks the end of the application programmer in favour of 'end-user' programming (Martin, 1982). Recent experience in the industry would indicate that this is not so, partly because prototyping contributes to an increase in user expectations. Weizenbaum (1976), for example, points out that the existence of any tool is pedagogical as well as practical, that is, people can see themselves changed with it (by imagining new things they might do). These extra expectations invariably outstrip the users' knowledge of the 4GL and require additional inputs from programmers with an in-depth understanding of this software tool. Another reason for the continued presence of the application programmer is that all prototyping tools are heavily machine and software based. The associated language, environment, and to a certain extent, culture of prototyping are those familiar to and comfortable for, the technologist (programmer/designer), and obviously much less well known to the user (even the confident spreadsheet-wielding male manager).

This last factor introduces a note of caution for those advocates of participatory design, and of human-centred design, not simply to embrace prototyping as *the* panacea for user control, but to recognize some of the inherent contradictions in this particular approach. In practice then, not only can prototyping be seen as within the technologists' arena, giving them the advantage of 'playing at home' but all possible answers (accepted prototype) must be machine-based, which means that other possibilities such as organizational change or manual procedures are rarely considered. In addition it must not be assumed that all members of the user-community are equally empowered, i.e. when it comes to evaluating a particular prototype then differences, such as gender, work-grade, organizational culture etc. will mean that some individuals are more likely to express an opinion (or remain silent) than others.

There are of course, many examples of positive methods to empower particular groups participating in systems development, such as study-circles (Vehvilainen, 1986; Bell *et al.*, 1988), but these are not well known to systems designers. This is not to propose that prototyping should be abandoned as a technique for participating in human-centred design, for it is clear that its 'real-world' images of computer screens, messages and printouts are a good deal more accessible to users than are the paper-based models (data-flow diagrams etc.) of structured methods. Indeed prototypes and mock-ups were successfully used in, and are advocated by, the Scandinavian Collective Resources Approach (see later sections). It is worth noting that some of the examples of prototypes developed in Scandinavia, (most notably the UTOPIA project in Sweden, Ehn, 1989) used paper-based mock-ups rather than machine-based ones, specifically in order to keep the debate on the terrain of the users.

Soft Systems Methodology (SSM)

This methodology has developed through use and reflection over the last twenty years or so, largely by students and staff at Lancaster University's Department of Systems and an associated consultancy company, ISCOL. The most significant explanations of this approach to 'systems engineering' are to be found in Checkland (1981) and Checkland and Scholes (1990). Like many other methodologies SSM was developed in response to the failures of 'hard' systems engineering, in particular where the complexity and diversity of human interests made the definition of a system's objectives difficult. SSM is essentially a 'managers' methodology, in that it is aimed at problem situations of concern to higher levels of management within organizations, such as strategy issues and decision support mechanisms. SSM has not been used to directly design computer-based information systems; indeed one of the methodology's strengths is that it does not assume computer systems are always 'the answer'.

Although SSM is not used in the arena we are most concerned with (the development of Information Systems within clerical work) some dimensions of this methodology are relevant to human-centred development. First and foremost is the shared understanding that in any work situation there is more than one perspective or *Weltanschauung* and that the objectives of any system will change according to the viewpoint adopted. Learning during the analysis stage is another facet of SSM and as Checkland (1981) notes, refers to a process of enquiry by the clients and problem solvers. SSM is also seen as a way to initiate debate between the various stakeholders: a discussion which has been termed 'unconstrained debate'. Critics of SSM, however, have been quick to point out that this can only profitably take place among participants with equal access to power and resources: unusual conditions in business organizations. In most situations the debate and participation encouraged by the methodology are vulnerable to manipulation by the more powerful stake holders (managers), supporting or reinforcing the *status quo*.

The developers of SSM are obviously aware of this problem, as the most recent description (Checkland and Scholes, 1990) of the methodology includes an explicit organizational analysis called 'a stream of cultural enquiry'. This comprises three elements: 1) an analysis of the roles of problem-solver, client and problem owner, 2) an analysis of the social system showing the norms and values associated with particular roles in the organization and 3) an analysis of the political-system, concentrating on the embodiments of power and the ways in which these are used within the organization under study. Checkland and Scholes do point out that the latter is exceptionally difficult to deal with; any intervention (i.e. use of SSM) will be part of the political processes within the organization, and the 'real' organizational politics will usually remain one step ahead (i.e. unwritten) of any expressed analysis. This probably explains why in all the case-studies given in their book, only one very short example of political analysis is given. Could it be that in practice the users (that is, the problem solvers) of SSM simply keep their political analysis at the level of unwritten thoughts and do not enter into any 'unconstrained debate' on these issues?

Certain elements of SSM have proved attractive to the developers of other methodologies; these have resulted in such hybrid approaches as multiview (Wood-Harper, Antill and Avison, 1985). Multiview is really a fusion of ideas from 'hard' methodologies (like SSADM) with features from SSM and socio-technical approaches (further discussion follows).

The Human Factors/Human Computer Interaction Approach

The human–computer interaction (HCI) approach has concentrated on the single user, giving an individual perspective to systems design. The concentration has been on the interface between the human user and the machine, aiming to produce more usable systems.

HCI has undoubtedly increased the focus on the human user of information systems and has been partly responsible for a shift in designers' perspectives towards bending the technology to meet the human's needs, rather than vice versa. Concepts of 'help' systems, for instance, have highlighted the idea of 'users' as being different, that is, not all having the same needs. A good example of this can be seen in some word processing systems, where experienced users are able to select a less intrusive level of 'help' information than would be required by a beginner using the same package. Much HCI research and development has established examples of, and principles for, good design of software interfaces (for example, Sellen and Nicol, 1990). These influences on designers have been far stronger in applications developed for powerful groups like middle and senior male managers, than in applications to be used by lower status workers (especially female employees). The exception to this trend comes in highly saleable, general purpose packages such as word processors where high turnover generates resources for high levels of research and development of the product. Even with this type of package, however, it is still common to see HCI researchers examining the use of this software in an academic environment, or a computer company's own research laboratories where the staff (atypically) have a lot of autonomy and are often male professional engineers. This is not to suggest that HCI ideas should not form part of human-centred design; however they need expansion since (on its own) the HCI approach tends to ignore the social and organizational context of which the system is a part.

A shift away from the emphasis on the 'individual' to an examination of the social and organizational can be detected within some more recent research work. These developments signal recognition that designed artefacts cannot be understood outside of the situations in which they are used (Suchman, 1987). A significant example takes the form of scenario-analysis (Carroll, 1990) a design mechanism which attempts to understand the (slightly) wider design context. Such new perspectives are also emerging in the development of CSCW (Computer Supported Cooperative Work) as an innovative area (Bannon *et al.*, 1988) and in projects like those carried out by Bødker and Gronbek (1991) using the Scandinavian tool perspective for the cooperative prototyping of multi-use–software. It is clear, however, from an analysis of the proceedings of the second European Conference on Computer-Supported Cooperative Work (Bannon *et al.*, 1991), that in practice, most of the CSCW studies are done with groups of professional engineers (often software engineers). This means that for the moment, one is more likely to encounter CSCW in a science-park than in a clerical-work environment.

Socio-Technical Systems Methods

Two of the leading socio-technical systems methods used in Britain are Mumford's ETHICS and HUSAT's (Human Sciences in Advanced Technology) user-centred design. The ETHICS method (Mumford, 1986) has been developed by Mumford and co-workers over a period of many years and has been referred to as participative design. In this method several design stages are gone through, with a participating design group which includes people on different job-grades. The method has mainly been documented by Mumford, and is in use in many organizations in the United States of America as well (Mumford, 1989). The HUSAT user-centred design method has been used with some success (Eason, 1987, 1988) by the HUSAT research centre. This method involves moving through several design stages, from feasibility to implementation. Both HUSAT's and Mumford's methods include an examination of job satisfaction and job design; HUSAT's method also researches the usability of the technology through human factors assessments. The use of such participative methods to overcome the problems of traditional design methods have been strongly advocated by many researchers into the effects of computerization (e.g. Hirschheim, 1985).

Significant criticisms of the above examples of socio-technical methods include the charges that they are not truly participative or democratic, are managerialist, and take too simplistic a view of job satisfaction, skill and the impact of technology (Ramsay, 1985; Ehn, 1989). Such approaches tend to take the technology as given and therefore concentrate on questions such as how to improve the usability of particular software rather than focus on contested explanations of what is defined as a 'technical' rather than a 'social/organizational' problem.

In practice, socio-technical approaches would seem at best to offer well-informed managers the opportunity to introduce computer systems with little overt disruption, at the price of investing staff time and convincing sceptical IT specialists. At worst, there is considerable potential for management to manipulate the situation by establishing their own selected design groups and bypassing the influence of workers' elected union representatives. These approaches, however, do have significant features of interest to human-centred development, in particular, that job design and organizational structures should be considered as part of systems design, and furthermore the understanding that much skill and expertise is held by the workers so their contribution to design is essential.

Human-Centred Development

Scandinavian Collective Resource Approaches to Systems Design

Although there are a clear set of factors in the political and industrial relations context that distinguish Scandinavia from Britain (see Ehn and Kyng, 1985), there is much of interest in these collective resource approaches to inform research into systems development, whether in the UK or elsewhere.

There are two significant and identifiable generations of research work to draw upon; the first, developed during the 70s and early 80s, focused on the formulation of strategies aimed at avoiding the fragmentation of work and deskilling that the implementation of new technology could bring. Some well-documented examples of these initiatives are: the NJMF project in Norway (involving the Norwegian Iron and Metal Workers Union), DEMOS in Sweden (Democratic Control and Planning in Working Life: On Computers, Trade Unions and Industrial Democracy) and DUE in Denmark (Democracy, Development and Electronic Data Processing); for further details see Ehn and Kyng (1985).

These collaborative research projects had several important dimensions: first, a recognition that the subject matter, the use and development of new technology and information systems, was multi-disciplinary, that is, that an understanding of the sociology of the organization was as important as a knowledge of technical possibilities; second, that the introduction of computing technology into the workplace could have a dramatic effect on the quality of jobs; third, if trade unions were to negotiate effectively on the introduction of new technology into an organization, then they would have to develop their own perspective on this technology. These influential projects produced results which included the following:

—data agreements (new technology agreements), allowing for employee access to information and for the election of data shop stewards (NJMF — Norway).

— development of a 'negotiation model for independent trade union investigatory work as a basis for participation in management projects' (DEMOS — Sweden).
— production of reports and teaching materials, based on surveys of local unions and the establishment of local working groups (DUE — Denmark).

These first-generation projects helped to strengthen the negotiating framework for trade unions in their dealings with employers over the introduction of new technology. The influence of these projects was significantly wider than just the workplaces involved; the academic researchers were able to disseminate the general principles from these particular experiences to a wide (international) audience. It was recognized by researchers and trade unionists that an intervention into the design of both technology and working practices was necessary, in order to influence the quality of jobs and services.

This led to the foundation of UTOPIA, the second generation research project, in 1981; here, computer scientists, social scientists and graphic workers were involved, with the support of the Nordic Graphic Workers' Union and of the Swedish Centre for Working Life (Arbetslievscentrum). The UTOPIA project aimed to develop a particular computer system (for newspaper page make-up and image processing) and associated working practices and also general principles for participatory design. The project encountered problems in disseminating its workstation design through conventional systems supplier channels; however, it also led to the formulation of some general principles in relation to systems design. These were summarized as follows by Ehn and Kyng (1985):

— Design of computer support is design of (conditions for) labour processes;
— Labour processes cannot be reduced to information processes;
— 'Use models' (user-oriented conceptual models) are important in design;
— Hardware should be considered early in the design, in parallel with software, not after;
— Important aspects of labour processes — in relation to design of computer support — cannot be formally described;
— Professional experience with and knowledge of the labour process are important in the design process;
— Professional experience with and knowledge of computers are important when designing computer support for a labour process;
— Design should be done with users, neither for nor by them;
— Mutual learning should be an important part of the work in a design group (before detailed work on specifications or design itself);
— 'Design by doing' is important, for example, simulating screen layouts and work processes with paper drawings and mock-ups. This enables 'tacit' or informal aspects of skills and processes to be included, and permits users to develop ideas without being constrained by (for instance) the technical limits of current prototyping approaches.

All of these research projects concentrated on workplaces dominated by men in strong craft unions (iron and steel-making, newsprint make-up etc.) and not on areas of work where women prevailed. Several smaller scale research projects have tried to redress the balance and also provide a gender perspective on systems design. Researchers like Vehvilainen (1986) and Olerup *et al.* (1985) stress the important of caring, communications and organizational skills in office work and argue that there is a tendency for these aspects of women's work to remain invisible to managers and systems designers. These researchers have concentrated on techniques to empower women at work to enable them to participate meaningfully in systems design. In this outlook, these feminist researchers have the notion of emancipatory practice in common with the larger scale projects (UTOPIA etc.) but they stress the importance of recognizing women's differing concerns and the different nature of their experience in relation to information technology.

Projects in Human-Centred Design

The Foundations (Within the UK)

As with the Scandinavian research projects described above, early human-centred developments in the UK were located in an area of work dominated by skilled male workers: manufacturing industry. A key initiative came from an organized group of trade unionists proposing alternative 'useful' work in the face of large-scale redundancies, the 'Lucas Aerospace Plan' (Cooley, 1987). A linked project, much influenced by the work of Cooley, came from an interdisciplinary research group who designed a human-centred lathe for small batch production within the engineering industry at the University of Manchester Institute of Science and Technology (UMIST; Rosenbrock, 1981, 1987).

The first of these developments, the 'Lucas Aerospace Plan' was an example of industrial workers being involved and interested in both the product — the notion of socially useful products rather than military ones — and the process through which products were being made. This second notion is linked to the ideas of retaining skilled work, fostering autonomy and improving the quality of working life through the design of new systems. The second 'project', the development of a human-centred lathe at UMIST, involved collaboration between engineers and social scientists; it also included Mike Cooley, one of the principal authors of the 'Lucas Aerospace Plan', on its steering committee. This project was based on the commitment to maintain (and enhance) human skill in the manufacturing process, in order to provide the flexibility and robustness not present in fully automated methods of production. An important concept underpinning this notion of skill-retention was that of tacit skill; that is, certain aspects of human skill may well be observed and appreciated in practice, but they cannot be completely abstracted, and are therefore not amenable to automation (Polanyi, 1962, 1957).

Continuing Human-Centred Work

A number of other important research projects have either arisen out of, or been strongly influenced by, the original work described above. A direct descendant of the UMIST work came in the form of a cross-European research project funded by ESPRIT, who invested £5,000,000 to develop a CIM (Computer Integrated Manufacturing) system in Britain, Germany and Denmark. This project attempted to incorporate the following principles:

—Technology must be designed in such a way that the synergy between human skill and computer power be optimized.
—Work within the factory should be organized so that in all areas people are able to apply their skills and have a significant degree of responsibility.
—Individual skill and competence should be increased through a balanced combination of learning by doing and formal training and education. (Rauner *et al.*, 1988)

There were a number of difficulties related to integration between both the research disciplines and the different countries involved, especially over the central issue of worker participation in the technological developments. Nevertheless, this project, like its counterparts in Scandinavia (such as UTOPIA), provided an innovative example which influenced researchers and designers in future work on the development of new IT systems.

Our own research work at Sheffield Hallam University in the UK (formerly Sheffield City Polytechnic), was strongly influenced by the original UMIST work, and is an interdisciplinary project in the area of human-centred office-systems. By extending the ideas of human-centred development into areas of office work where the workers are mainly women, we added the concept of gender relations to the already expressed notions of skill-retention, empowerment and design by doing. The results of our project, which is more fully described in Chapter 8, included the formulation of key design principles as set out below.

Finally, another human-centred research project which goes beyond the manufacturing arena is the 'Parosi' project carried out at Brighton Polytechnic (which is now the University of Brighton); Gill, 1990. Supported by the EC Social Fund, this project brought together health workers, computer scientists and women from Asian communities to construct a knowledge-based system for providing advice on nutrition. It is described by Gill as having helped to bring the focus of diversity into human-centred systems research, whether in terms of cultures, of values or of disciplines. While this could help to create a basis for drawing gender issues into the human-centred systems framework, these are not an explicit part of the 'diversity' Gill describes. His analyses of the project do not address gender as a significant issue in interactions between the different parties involved, or in the priorities which they identified.

Our Human-Centred Perspective

In developing our own perspectives on human-centred design, we have incorporated many of the ideas described above. Our practical work has been heavily influenced by the following triad of design concepts: design by doing, designing for skill and empowering.

Human-centred design is among the plethora of design alternatives which appealed to us for many reasons. It enabled us to break away from the narrow-minded perspectives of traditional methods, which aim for objectivity and 'scientific' authority, and result in abstracting the system design from both social and work contexts. Similarly, although human-centred design builds upon concepts derived from social science methods (for example, socio-technical, HCI), it recognizes the social and political aspects of the design process and their relationship with technology.

Human-centred design adopts a different philosophical position as its basis, one which incorporates tacitness, empowering, culture and social values. System design is therefore not abstracted from its social context of use but becomes an integral part of it.

The researcher/system designer involved in human-centred design inevitably becomes part of the social context. In a similar way the designer when involved in design by doing, designing for skill and the empowering process becomes partisan, supporting those who are being empowered and whose working practices and conditions are to be changed. The human-centred designer must be prepared to fully engage in the social and organizational context of the design situation. Only through such engagement can the designer come to understand and learn about the users, their skills, and their system requirements. Through such understanding the designer can facilitate the design process, helping the participants to realize their new roles, and to specify, choose and implement a new system. For all concerned it necessarily becomes an intimate process of change and learning.

It is important to stress that we do not see human-centred design as a static method. We view it as a dynamic approach which can only be developed through practice. Unlike the traditional methods described above which have spawned many texts, we do not envisage that there can be a 'how to' textbook on human-centred design, a view shared with Greenbaum and Kyng (1991). With human-centred design, the process does not start at stage one and finish at stage ten; instead the human-centred design process is both continous and situationally based. This perspective offers a design philosophy and theory, with an accompanying set of examples which can and must be adapted for new contexts.

Outlined below are the key, general principles of human-centred design which we have derived from the case-study research (described in its own right in Chapter 8). It is our view that these principles state the major necessary conditions for the performing of a human-centred design process. They are not intended as an exhaustive list, but as a helpful framework for others attempting to develop human-centred approaches.

Principles of Human-Centred Approaches to Systems Development

1 The introduction (development) of a new system should focus on the organizational changes, user needs and demands *together* with the technological requirements.
2 The boundaries between which issues are defined as 'technical' and 'organizational' are not fixed and need to be negotiated.
3 In most cases 'new' applications of technology should be seen as the development of permanent support systems (that is, they become part of the life of the organization) not one-off projects which finish with implementation.
4 Humans should be seen as the most important facets of an information system and should be 'designed in', helping to provide those qualities of correctness, robustness and extendibility so sought after by software engineers.
5 The people-context of information systems must be studied and understood, for it is clear that dimensions such as gender, race, class, power,

career paths etc. are significant aspects of a work-force/organization. Design can only proceed from an informative analysis of these factors.

6 The development process should allow for cross-over primarily of an understanding (learning) between applicative knowledge and computing knowledge and secondly for career interchange.

7 In a development that seeks to replace/enhance existing information systems, it must be recognized that much skill and knowledge possessed by the work-force is tacit and often makes a significant contribution to the smooth running of an information system. Designing for skill, therefore, is a necessity.

8 Design involvement/participation should be active and contain some form of design by doing, for example, role playing situations, talking through particular work scenarios, drawing up new job descriptions and career structures.

9 Participation of the various interest groups is essential. These interest groups could be based on several categories: levels of management and other work grades, men/women, functional area of work and representative organizations such as trade unions.

10 Participation has to be worked at. The participants need to be 'empowered' through access to information, training and new forms of organizational structure and communication.

11 Consideration must be given to the most visible parts of a computer system: the human–machine interface. However, the work-force should be recognized as social groups and not merely be seen as individuals interacting with a computer.

12 The organizational context of the information system must be fully understood, (not just the functional interactions typically reproduced in systems development 'data flow diagrams'). Recognition of common objectives, like quality of service provision, and of common constraints such as external budgets, can help to facilitate negotiations between different interest groups.

Conclusion

Although there will be continued development and introduction of 'new' scientific approaches to systems development, it is clear that many of today's techniques and principles will be with us for many years to come. A good illustration of this point is provided by object-oriented analysis (OOA) and design, seemingly very much the flavour of the early 1990s but based on the programming ideas of the 70s. Coad and Yourdon (1991a) exemplify this cumulative approach in the following statement:

> What we are attempting to do is to incorporate the best ideas of the first three methods (functional decomposition, data-flow approach, information modelling) in a more comprehensive, all-encompassing method — OOA.

Although we would agree with Friedman and Cornford (1989) that the current phase of computer systems development is characterized by an emphasis upon 'constraint by user relations', it is clear that many of the development methods and tools being used are those developed to solve the problems of software reliability, software productivity and software applicability of earlier times. In practice then, the actual methods followed tend to concentrate power and decision-making in the hands of the technologists.

If human-centred developments are to become more commonplace, then, it is clear that this will require organizations to adopt a practice of giving priority to user requirements and organizational needs over technological design methods. Friedman and Cornford indicate the kinds of strategies organizations are using to allow their computer systems developers to become more concerned with user relations, for example, project teams dedicated to particular users, devolving the information systems function to user departments, and schemes to involve users in systems development. We see the second of these as being the most fruitful for human-centred developments in that it would allow a much closer identification (by staff) of the relationships between technological and organizational issues. It would also allow the computer systems to be seen as long term support systems for the organization/department, rather than one-off development projects. This 'project' style of work is one that militates against human-centred developments; although it can permit certain levels of user participation and control, its short term emphasis accentuates the split between the knowledgeable users and the technologists, undermining the motivation for users to learn of the technology and for information systems people to properly understand the application. Human-centred development methods tend to highlight the different aspirations of particular interest groups, but the sharing of common objectives such as 'the quality of service provision' and common constraints such as external budgets, can help in reaching negotiated positions.

In order to reach such negotiated positions, bridges of understanding and respect have to be built between the different interest groups — the technologists and users, the workers and managers. All parties need to be recognized as having a legitimate position and be regarded as 'experts' in their own areas. Human-centred design, therefore, requires different ways of thinking about technology: its use, participation, and organizational functioning.

It is understandable that organizations should be attracted to traditional formal computing methods. Such methods present themselves as being straightforward in technique, as scientific and objective, and of having a body of professional and certificated experts (system analysts) to carry them out. They also stress the 'gee whiz' wonders of technology and its intrinsic ability to improve efficiency, productivity and profitability. All of this gives them an impressive face value. However, as Leith (1990) points out system design can be more art than science and the increasing reports of system failures can only lead to a healthy scepticism by users and system procurers. No longer can such failures be placed squarely at the door of human error or user ineptitude.

Human-centred design embraces organizational, work group, and individual complexity, offering participative and negotiated methods of selecting alternative development strategies, which need not be technological. Social influences will obviously continue to be important and it is clear that regional differences will endure; for example, the European Community with its penchant for policy-making contrasts strongly with the US 'market winner' approach. Within the European Community it is clear that the more participative and human-centred design methods will find a more comfortable home in countries such as Germany and Denmark with their strong democracy at work cultures, rather than in the 'market forces' climate currently prevailing in Britain.

It is to be hoped that some of the European human-centred research outlined above will influence future collaborative developments in design methods, especially with the expansion of the EC and the associated harmonization of methods of systems development across Europe. The initial signs are not promising, however, as illustrated for example by the deliverables (documents) produced by the 'Euromethod' project. It appears that the large (economic) effort behind 'Euromethod' is being dominated by the likes of structured methods such as SSADM and their sponsors, British Telecom, and that the emphasis will be on techniques for doing design and methods of project management. In the short term, it looks very unlikely that enlightened aspects of participatory design, human-centred design or of CSCW will figure strongly in Euromethod, but will remain instead under the token heading of 'Importance of the human being' (Euromethod Project, 1991).

This situation yet again reduces the human beings involved to marginal status. However, there are optimistic voices. As Greenbaum and Kyng (1991) comment:

> Throughout the Western world, democracy at work is clearly the next frontier . . . the field of systems development needs to adapt to this changing environment.

References

BANNON, L., BJORN-ANDERSEN, N. and DUE-THOMSON, B. (1988) 'Computer Support for Cooperative Work: An Appraisal and Critique' in BULLINGER, H.J. *et al.* (Eds) *Proceedings of the First European Conference on Information Technology for Organisational Systems*, Amsterdam, Elsevier.

BANNON, L., ROBINSON, M. and SCHMIDT, K. (Eds) (1991) *Proceedings of the Second European Conference on Computer-Supported Cooperative Work*, Dordrecht, Kluwer.

BELL, D., GREEN, G., MURRAY, F., OWEN, J. and PAIN, D. (1988) 'Human centred design — A methodology for action', in BULLINGER, H.J. *et al.* (Eds) *First European Conference on Information Technology for Organisational Systems*, Amsterdam, Elsevier.

BØDKER, S. and GRONBEK, K. (1991) 'Cooperative prototyping: Users and designers in mutual activity', *The International Journal of Man-Machine Studies*, **34**.

CARROLL, J.M. (1990) 'Infinite Detail and Emulation in an Ontologically Minimized HCI', in CHOW, J.C. and WHITESIDE, J. (Eds) *Proceedings of CHI*, New York, ACM.

CHECKLAND, P. (1981) *Systems Thinking, Systems Practice*, Chichester, Wiley.

CHECKLAND, P.B. and SCHOLES, J. (1990) *Soft Systems Methodology in Action*, Chichester, Wiley.

COAD, P. and YOURDON, E. (1991a) *Object-Oriented Analysis*, (2nd Edn) Englewood Cliffs, NJ, Prentice Hall.

COAD, P. and YOURDON, E. (1991b) *Objected-Oriented Design*, Englewood Cliffs, NJ, Prentice Hall.

COMPUTING (1991) *Euromethods*, London, 17 October.

COOLEY, M. (1987) *Architect or Bee? The Human Role of Technologies*, London, Hogarth.

CUTTS, G. (1987) *Structured Systems Analysis and Design Methodology*, London, Paradigm.

DAMODARAN, L., IP, K. and BECK, M. (1987) 'Integrating Human Factors Principles into a Structured Design Methodology', in BULLINGER, H.J. *et al.* (Eds) (1988) *Proceedings of the First European Conference on Information Technology for Organisational Systems*, Amsterdam, Elsevier.

DEARNLEY, P. and MAYHEW, P. (1983) 'In favour of system prototypes and their integration into the systems development life cycle' *The Computer Journal*, **26**(1).

DEMARCO, T. (1978) *Structured Analysis and Systems Specification*, Englewood Cliffs, NJ, Prentice Hall.

EASON, K.D. (1987) 'Methods of planning the electronic work place', *Behaviour and Information Technology*, **6**, pp. 229–38.

EASON, K.D. (1988) *Information Technology and Organizational Change*, London, Taylor and Francis.

EHN, P. (1989) *Work Oriented Design of Computer Artifacts*, Hillsdale, NJ, Erlbaum.

EHN, P. and KYNG, M. (1985) 'Trade unions and computers: The Scandinavian Collective Resources Approach', paper presented at the FAST conference *The Press and the New Technologies — the Challenge of the New Knowledge*, Brussels, 7–9 November.

EUROMETHOD PROJECT (1991) Phase 2, Deliverable 1, *State of the Art Report*, Issue 2.

FLOYD, C. (1984) 'A systematic look at prototyping' in BUDDE, R., KUHLENKAMP, K., MATHIASSEN, L. and ZULLIGHOVEN, H. (Eds) *Approaches To Prototyping*, Heidelberg, Springer-Verlag.

FRIEDMAN, A.L. and CORNFORD, D.S. (1989) *Computer Systems Development: History, Organization and Implementation*, Chichester, Wiley.

GILL, K.S. (1990) *Summary of Human Centred Research in Europe*, research paper, SEAKE Centre, Brighton Polytechnic.

GREEN, E., OWEN, J. and PAIN, D. (Forthcoming) 'Designing Systems, Defining Roles' in WOOLGAR, S. and MURRAY, F. (Eds), *Social Perspectives on Software*, MIT Press.

GREENBAUM, J. and KYNG, M. (1991) *Design at Work: Cooperative Design of Computer Systems*, Hillsdale, NJ, Lawrence Erlbaum.

HIRSCHHIEM, R.A. (1985) *Office Automation: A Social and Organisational Perspective*, Chichester, Wiley.

JOHNSON, T.J. (1972) *Professions and Power*, London, Macmillan.

LEITH, P. (1990) *Formalism in AI and Computer Science*, Chichester, Ellis Horwood Ltd.

MARTIN, J. (1982) *Application Developments without Programmers*, Englewood Cliffs, NJ, Prentice Hall.

MEYER, B. (1988) *Object-Oriented Software Construction*, Englewood Cliffs, NJ, Prentice Hall.

MUMFORD, E. (1986) *Using Computers for Business Success: The ETHICS Method*, Manchester, Manchester Business School Publications.

MUMFORD, E. (1989) 'User Participation in a Changing Environment — Why We Need it' in KNIGHT, K. (Ed.) *Participation in Systems Development*, London, Kogan Press.

OLERUP, A., SCHNEIDER, L. and MONOD, E. (Eds) (1985) *Women, Work and Computerization: Opportunities and Disadvantages*, Amsterdam, Holland.

OWEN, J. (1992) 'Women Developing Office Systems: Case-studies in User-Involvement During Design', PhD thesis, Sheffield City Polytechnic.

POLANYI, M. (1957) *Personal Knowledge*, London, Routledge and Kegan Paul.

POLANYI, M. (1962) 'Tacit knowing: Its bearing on some problems of philosophy', *Review of Modern Physics*, **34**, pp. 601–5.

RAUNER, F. *et al.* (1988) 'The social shaping of work and technology', *AI and Society*, **2**, 1, February, p. 47.

RAMSAY, H. (1985) 'What is participation for? A critical evaluation of "labour process" analyses of job reform', in KNIGHTS, D., WILLMOT, H. and COLLINSON, D. (Eds) *Job Redesign — Critical Perspectives on the Labour Process*, Aldershot, Gower Publishing.

ROSENBROCK, H. (Ed.) (1981) *New Technology: Society, Employment and Skill*. London, CSS (Council for Science and Society).

ROSENBROCK, H. (1987) 'Technology and society', paper given as Cockcroft Lecture, 1986, *Manchester Technology Association*.

SELLEN, A. and NICOL, A. (1990) 'Building user-centred on-line Help', in LAUREL, B. (Ed.) *The Art of Human–Computer Interface Design*, Reading, MA, Addison-Wesley, pp. 143–53.

SUCHMAN, L. (1987) *Plans and Situated Actions: The Problem of Human-Machine Communication*, Cambridge, Cambridge University Press.

VEHVILAINEN, M. (1986) 'A study circle approach as a method for women to develop their work and computer systems', paper for IFIP Conference, *Women, Work and Computerisation*, Dublin, August.

WEIZENBAUM, J. (1976) *Computer Power and Human Reason: From Judgement to Calculation*, (1st Edition) London, Pelican Books.

WOOD-HARPER, A., ANTILL, L. and AVISON, D. (1985) *Information Systems Definition: The Multiview Approach*, Oxford, Blackwell.

YOURDON, E. (1986) *Managing the Structural Techniques*, New York, Yourdon Press.

Chapter 2

Establishing Gender Perspectives on Information Technology: Problems, Issues and Opportunities

Flis Henwood

Introduction

There is, now, a substantial body of literature examining the relationship between women (and sometimes men) and technology. This 'gender and technology' literature has developed particularly rapidly over the last decade, starting in the early 1980s with research into the effects of 'new technologies' on women's jobs, and developing, in recent years, into a wide ranging set of debates about the relationship between gender and technology.[1] How far have we come in our thinking? What has been learnt so far? What problems remain? How should we proceed? These are the questions that underlie this chapter which aims to address these issues in relation to the specific area of computing/information technology (IT).

I am particularly concerned with the relationship between theory, the concepts and frameworks with which we seek to understand and explain the gender and IT relationship, and change, the strategies we develop to enable us to transform the present gendered relations of technology. In reviewing the literature, I have identified two very clear and different approaches to the development of gender perspectives of IT. The first, which I have termed the 'women in technology' approach, focuses almost exclusively on women's exclusion from technological work, with change understood as coming about via increased access and further equal opportunities policies. The second approach is characterized by a broader focus: examining the nature of technological work, its development over time and its articulation with changing gender relations. A key issue for this 'women and technology' approach has been the gendering of technological skills and knowledges.

What has each of these approaches contributed to the gender and IT debate? How do the two approaches differ and what are the implications of each approach for change? In seeking to provide answers to these questions, I will be paying particular attention to the ways in which the two key concepts, 'gender' and 'technology', are articulated within the feminist literature. It is my contention that the continued existence of elements of both biological and technological determinism (even where conscious effort has been made to avoid these forms of determinism), is seriously inhibiting the development of appropriate transformation strategies. Towards the end of the chapter, I suggest some ways in which the limitations of existing frameworks might best be overcome; and how, by moving towards the construction of a new framework for understanding gender and IT, we will be in a stronger position for developing both meaningful and effective strategies for changing that relationship.

Women in Technology: A Question of Access and Equality?

Women's exclusion from technological work has always been a major focus in research on gender and technology. Studies of women and engineering, for example, have drawn attention to women's under-representation in technical occupations such as technician and professional engineer, and their over-representation in operator and clerical jobs (Cockburn, 1983; EITB, 1984; Henwood, 1991a and 1991b). More recently, equivalent studies have been undertaken to assess the extent to which such patterns of gender segregation are discernible in the computing/IT sector.

A study of the US computer industry (defined in terms of both the hardware and the software sectors), showed women concentrated in the lowest level occupations in both sectors (Strober and Arnold, 1987). Figures for 1980 show that, in software, while women represented 92 per cent of data entry workers and 59 per cent of computer operators, they were a much smaller proportion of programmers and of computer scientists and systems analysts (31 and 22 per cent, respectively). Similarly, on the hardware side, women were 72 per cent of all assemblers but only 15 per cent of electronics technicians and 5 per cent of engineers.

While this position was, in some occupational categories (computer science/ systems analysis, programming, and even technician and engineering jobs), an improvement on the 1970 figures, the research shows that during a period when computer-related occupations in the US grew by 80 per cent overall, gender segregation continued. Indeed, it is significant that while some women are making some inroads into technical and higher level occupations, there are, at the same time, signs of increasing feminization of some of the lower level jobs. Thus, between 1970 and 1980, women's representation increased from 90 to 92 per cent and 29 to 59 per cent in data entry and computer operator jobs, respectively (Strober and Arnold, 1987, p. 146).

The picture changes slightly when we examine more recent data. Table 2.1 adds 1990 data to Strober and Arnold's 1970 and 1980 data. It shows clearly that, during the 1980s, women continued to increase their share of all computer scientist/ systems analysts and programming jobs, (to 34 and 36 per cent, respectively) and while they have continued to increase their share of the lower level operator jobs, (a 66 per cent female occupation by 1990), their share of the lowest level data entry jobs has fallen to 87 per cent of the total. On the hardware side, too, women's share of assembly jobs continued to decline, decreasing by 6 per cent between 1980 and 1990, while their share of engineering jobs rose by 5 per cent. These figures give cause for some optimism in that they suggest that there has been a slowing down in the process of gender segregation in computing and computer related occupations in the 1980s.

Table 2.1: Women as a proportion of all employees in selected computer-related occupations 1970–1990, USA (%)

	1970	1980	1990
'Software'			
Computer Scientists and Systems Analysts	15	22	34
Programmers	23	31	36
Computer Operators	29	59	66
Data Entry Clerks	90	92	87
'Hardware'			
(all electrical and electronic engineering)			
Assemblers	74	72	66
Technicians	11	15	15
Engineers	2	4	9

Note: All figures are rounded up to nearest whole number.
Source: 1970 and 1980 figures taken from Strober and Arnold (1987); 1990 figures were calculated from unpublished tabulations from US Department of Labor, Bureau of Labor Statistics, 1990 Annual Averages.

Another US study, this time of the 'software industry' (Kraft and Dubnoff, 1983), also analyzes the extent of gender segregation in computing. While this study shows that women are found in the software industry and are better represented here than on the hardware side (in the early 1980s, women were 24 per cent of the 'software workforce' in the US, compared with 2 per cent in electrical engineering), it also provides some very valuable data concerning women's segregation within the software industry. In particular, women in this study were concentrated in the lowest paid jobs in the industry, in business programming, documentation and maintenance, as well as in the worst paying industries, in financial, real estate and communications. The study also found that while 22 per cent of women in the sample were managers of some kind, men were three times more likely to be upper level managers than women (10 per cent of men compared with 3 per cent of women). Furthermore, women managers were found to be more likely to manage other women whereas men tend to manage men and women, causing Kraft and Dubnoff to argue that the status of managers is given, in part, by the status of the people they manage.

In addition to these gender divisions, we find that the lowest level software jobs are populated by racial and ethnic minorities. Strober and Arnold found that whereas white men were over-represented in computer science/systems analyst and computer programmer categories and under-represented in computer operator and data entry jobs, with the exact reverse being true for white women, black men were under-represented in the top two categories and black women were more heavily concentrated in low level computer-related occupations than white women (Strober and Arnold, 1987, pp. 150–2). Table 2.2 shows that this overall picture holds true for 1990, although black women are increasing their representation in the higher level computer scientist and systems analyst jobs, as well as in the lower level operator and data entry jobs where they continue to be over-represented. This table also demonstrates clearly the importance of such racial breakdowns for understanding the position of women in computing. It appears

Table 2.2: White and black workers as a proportion of all employees in selected computer-related occupations, by gender 1980–1990, USA (%)

	Total Labor Force		Computer Scientists and Systems Analysts		Programmers		Computer Operators		Data Entry Clerks	
	1980	1990	1980	1990	1980	1990	1980	1990	1980	1990
White										
Men:	50.0	47.8	71.3	57.1	62.2	56.3	33.9	28.6	5.7	9.4
Women:	36.0	38.7	19.5	29.6	26.5	30.6	49.5	54.8	71.3	63.5
Black										
Men:	4.8	5.0	3.0	3.8	3.1	3.4	4.7	4.2	1.1	2.5
Women	4.8	5.1	1.7	3.1	2.5	2.4	6.9	8.7	15.1	17.1
Other	4.4	3.4	4.5	6.3	5.7	7.3	5.0	3.7	6.8	5.2
TOTAL:	100	100	100	100	100	100	100	100	100	100

Note: All totals are rounded up to nearest whole number.
Source: 1980 Figures taken from Strober and Arnold (1987); 1990 figures calculated from unpublished tabulations from US Department of Labor, Bureau of Labor statistics, 1990 Annual Averages.

that the 'defeminization' of data entry work, shown in Table 2.1, only holds true for white women; black women's share of this occupation continues to rise, along with (all) men's.

Despite the lack of comparable research and data on the UK computer industry (particularly on race/ethnicity), preliminary assessments suggest that a similar pattern of segregation to that in the US exists here. A survey undertaken in 1986, for example, found that women represented 2 per cent of data processing managers, 12 per cent of programmers and 95 per cent of data preparation staff.[2] Certainly, educational data shows a distinct gender pattern with women representing a small, and declining, proportion of entrants to university computer studies courses. UCCA university entrance figures show women representing 24 per cent of entrants in 1980 but only 10 per cent in 1987 (Dain, 1988; Lovegrove and Hall, 1987). Many factors may be responsible for this worsening position of women in computer higher education, but one commentator at least has suggested that the introduction of microcomputers into secondary schools may be largely responsible (Newton, 1991, p. 145). Certainly, there is much evidence to support the notion that the way in which computers have been introduced into schools — the pupil computer ratio, the location (in science and maths classes), and the emphasis on experience gained in 'computer clubs' — has benefited boys rather than girls (Newton, 1991; Culley, 1986; Hoyles, 1988). However, the most interesting questions of all have yet to be asked, let alone answered. How, exactly do these (and other) factors combine to produce a masculine computer culture in schools? Is there just one or several different cultures coexisting in schools? What is it about masculine cultures that deter girls and women? What would girls and women want to see in place of the existing culture? These are central issues to which I shall return at the end of the chapter.

Women in Technology research has focused on factors other than exclusion and/or segregation, however. Conditions of work for women in technology has also been a focus of concern. In relation to the IT sector in particular, attention has been drawn to women's and men's differential salaries. Strober and Arnold found that women's annual and hourly earnings were substantially less than men's: 72 per cent of men's on average in 1980, after controlling for industry, age, education and experience (Strober and Arnold, 1987) and in Kraft and Dubnoff's sample, women's salaries were found to be 70 per cent of men's on average and, after controlling for age, education and experience, they were still only 85 per cent of men's. Furthermore, when size of organization was taken into account, their figures show that women in small organizations are even worse off, often earning barely 50 per cent of men's salaries (Kraft and Dubnoff, 1983). A UK study of women computer professionals undertaken by the British Computer Society (BCS) showed that the majority of women had salaries within the £16–£20,000 range and that there appeared to be a £25,000 'ceiling' for women, that did not exist for men. The study also found that having a career break has the effect of reducing a woman's salary by between £5,000 and £10,000 per annum (Beech, 1990).[3]

Studies such as these have drawn attention to the important fact that gender segregation patterns in this relatively new industry parallel, to a large extent, those found in the more traditional industries. In the IT sector, women are not, as yet, gaining access to the higher status jobs in systems design and management at anything like the same rate as men. For many, this is seen as a problem of unequal opportunities and there has therefore been much support for the recent Women into IT (WIT) campaign which has industry backing for a set of initiatives designed to encourage women into IT careers.

For many feminists working within this 'women in technology' framework, increasing the numbers and proportion of women in computing/IT is seen as the solution to the problem. Equality for women and men in IT is seen as coming about via increased access for women into IT jobs and careers; support for campaigns such as WIT is therefore a central feature of their strategy for change.[4] However, this position has been criticized by other feminists, including myself, who see the women and technology issue as being about far more than increased access to high technology jobs. Among this group, there is scepticism of campaigns such as WIT,[5] not least because of the way in which these initiatives have been linked to the perceived skills shortages within the IT sector. This scepticism seems entirely justifiable when we consider how the WIT campaign explains its origins:

> The need for a campaign emerged at the beginning of 1988 when an analysis of the demographic trends highlighted the fact that shortages of skilled staff are by far and away the most significant obstacle to the more widespread and effective use of IT now and for the future. (WIT, 1989, p. 1)

and its key objectives:

> To help employers to overcome current and prospective shortages of IT skilled staff by raising the number and proportion of girls and women entering and sustaining IT related careers at all levels. (WIT, 1989, p. 1)

Employers, too, seem clear about their reasons for supporting such initiatives:

> a 1 per cent reduction in staff turnover saves £500,000 off my bottom line and our experience is that women are indeed more loyal if you treat them properly. (WIT, 1990)

and

> we attach great importance to this campaign as part of the joint Government–Industry work in overcoming the great IT skills shortage in this country. (WIT, 1990)

The fact that campaigns such as WIT are tied in closely with the needs of business and, for employers at least, are not usually conceived within the spirit of equal opportunities, has led many feminists to question the extent to which such initiatives can be anything other than a very small part of the solution to the women and technology 'problem'. However, as Hacker has shown, even where such initiatives are given industry support as equal opportunities initiatives, there are convincing reasons to remain critical of such initiatives.

In an interesting account of some action-research she undertook in the early 1970s at AT&T (Bell Company) (Hacker, 1989; Smith and Turner, 1990) Hacker discusses her team's efforts to promote 'an even distribution of women and minorities from top to bottom of the corporation', including technical occupations (Hacker, 1989, p. 19). She recounts that it was only later, as the research progressed, that she and her colleagues realized that 'more women were going to be moved out than up' (Hacker, 1989). Reflecting on these early experiences, Hacker writes, 'Belatedly, we had discovered the process of technological displacement and that it affects women differently, at different times, than it does men' (1989). She points out how some changes that she and her colleagues were arguing for, conceived as they were in the spirit of 'equal opportunities', were lost because they took no account of the company's strategy for organizational and technological change. For example, arguments for the restructuring of the telephone operator's job (to increase job satisfaction and thereby reduce staff turnover) were lost precisely because high turnover was part of the AT&T's automation strategy — enabling them to automate the operators' jobs without layoffs (Hacker, 1989).

Other elements of the AT&T action-research played right into the hands of the company. As Hacker writes, 'We found ourselves in a position of arguing that women and minorities be hired exactly where the company wanted them — in jobs next to be automated' (pp. 21–22). Again, the liberal framework, conceived around the notion of 'equality' and 'equal opportunities' proved inappropriate. Moving women and minorities into, and higher within, the organization was simply not enough. The AT&T research demonstrated the need to understand the role of technology and technological change as part of the company's overall strategy for reducing labour costs and AT&T as 'a giant corporation within a capitalist system' (Hacker, 1989, p. 24).[6]

Hacker's story points to some very serious criticisms that can be made of the equality/women in technology approach. In particular, we can note that the equality approach takes technology as 'given'. It accepts that with any given technology, certain 'effects' are inevitable, including the need for specific skills. Anyone able to acquire and offer these skills will therefore be in demand. It follows, then, that women wanting access to IT careers must acquire these skills. There are clear elements of technological determinism within this position. The technology is seen as unproblematic, as is the notion of 'skill', which derives from it. However, as Hacker's research makes clear, the process of technological change is far from neutral and access to skills is not enough to secure women a permanent place in technological work.

With regard to gender, we can see that the equality approach takes an essentialist position. In seeking to understand more about how the equality approach understands 'women' and how it perceives the problem of gaining equality for women, Fox-Keller's work on gender and science (1986) is useful. In examining the equality approach in debates on gender and science. Fox-Keller determines from whence it comes. She wants to understand, in particular, why those who support the equality approach seek equality for women through the provision of the same opportunities as men. This position, she argues, can be understood as a reaction to the emphasis on difference in dominant discourse, which supporters of the equality approach claim has served them badly. She argues that many women take the position that the division of the world into a series of dichotomies (masculine/feminine, rational/irrational, mind/body etc.) has led either to their exclusion from, or to their segregation within, scientific/technological work. It is in response to this, then, that some women have claimed 'we are not different' (Fox-Keller, 1986, p. 168). This position, argues Fox-Keller, rejects the notion that people are divided into two according to their sex, and asserts instead, the unitary nature of humanity (her 'two-one step').

The classic liberal feminist position found within the 'equality' or 'women in technology' approach embodies strong elements of a liberal-humanist notion of 'the individual': an autonomous, rational person existing ontologically, if not empirically, outside of the social context (Jagger, 1983, p. 28). Notwithstanding physiological differences, men and women are understood as essentially equal. Any inequality that exists between the sexes has been 'added on' by society and can, it follows, be 'peeled away' to reveal the 'natural' state of equality and unity of the sexes. From this perspective, equal opportunities programmes simply give women back what they have essentially — equal status with men.[7] However, as Fox-Keller makes clear, women cannot gain equality by demanding to be treated the same as men. The problem lies in the fact that 'men are not neutral people'; the universal 'man' is, of course, male. Both women and men are gendered, and arguing for women to be given the same as men will not result in equality for women. Rather, it will lead to invisibility and unrepresentedness, what Fox-Keller refers to as being 'negated in the quest for assimilation' (1986, p. 169).

Finding the equality approach unsatisfactory, many feminist commentators have sought to shift the focus of concern away from women in technology, towards what they consider to be more relevant questions and issues. How did it come about that women and men hold the positions they do in relation to technology and technological work? How have these positions changed over time and what factors are responsible for these changes? An interest in these issues led to a focus on women (or, increasingly, gender) and technology which examines the historical development of women's (and men's) participation in technological work, and on the questions concerning the culture of technological work and the technological workplace. It is to this literature, particularly as it relates to computing/IT, that I will now turn.

Women and Technology: The Gendering of Technological Skills and Knowledges

Feminist work on the history of computing/IT has drawn attention to the fact that, since Ada Lovelace first worked with Charles Babbage on his Analytical Engine,[8] women have always been involved in computing. Grace Hopper's work on early compilers and the development of COBOL language is well-known and often noted and it is also widely acknowledged that women were centrally involved in designing software for ENIAC[9] in the USA in the 1940s (Griffiths, 1988; Shurkin, 1985; Perry and Greber, 1990). However, as several commentators have pointed out, early 'programming' was defined as clerical work and not considered a very high status occupation (Kraft, 1979; Strober and Arnold, 1987). Indeed, it is to this relationship between changing occupational structures and changing skill definitions that we have to look to understand the history of the relationship between women (or gender) and computing.

Only slowly in the development of computing did it become clear that 'programming' involved a knowledge of logic, maths and electronic circuits, and it was then that it became redefined as a high-level, challenging and creative occupation. As both hardware and programming techniques were developed in the late 1950s and 1960s, programming became defined as a skilled occupation, what Kraft (1979) calls a 'craft' occupation, and at this point, men joined the ranks of programmers (although, throughout the 1960s and early 1970s, programming continued to be an important source of employment for women, too). As software design techniques developed further during the 1970s, there developed a division of labour within programming, with overall design elements (specifications etc.) being separated off from the more routine tasks of writing the programs.[10] With this technical division of labour, there developed the sexual division of labour that we see now, with women dominating in the lower echelons of the computer hierarchy and men dominating in the upper echelons.

Some very useful early work on the computer industry sought to explain the changing structure of the industry and women's position within it in the context of labour process theory. According to Kraft (1979) and Greenbaum (1979) for example, women entered computing only to work in jobs that were either defined as low in skill from the start (early programming) or had recently been created as a result of the fragmentation and deskilling process taking place in the industry as a whole (as programmers, computer operators and data entry clerks). Following Braverman (1974), Kraft and Greenbaum both analyze technological change within the context of the capitalist labour process, with an emphasis on management control and deskilling and like Braverman, neither take a particularly critical look at the relationship between skill and gender, a project that has since been taken up by many feminists working on gender and work/technology issues. As Cockburn makes clear in her now well-known study of gender and technology in the print industry (Cockburn, 1983), the relationship between technological change, deskilling and the entry of women as conceived in Braverman-type labour process theory, is undermined by the empirical evidence which suggests that skill levels are not determined by the technology alone (or even at all), but by the relative negotiating strength and power of different actors in the workplace.[11] Thus, male compositors managed to negotiate with management to retain their skilled craft status despite the introduction of new computerized technology which required the 'feminine' skill of keyboarding.

Game and Pringle similarly build upon traditional labour process theory in their analysis of the social construction of skill categories, and the gender dimension within this (Game and Pringle, 1984). In their series of case-studies of gender and the labour process in Australian manufacturing and service industries, they too show how skill is an ideological and social category, not merely a technical one. In the whitegoods industry, for example, they show how male workers resist the 'feminization' of work, which they fear will result from the introduction of new technologies, by creating and supporting a whole series of dichotomies which (they argue) distinguish between men's and women's work. According to Game and Pringle, distinctions such as heavy/light, dirty/clean, technical/non-technical, mobile/immobile are constructed to preserve a sexual division of labour, the whole basis of which is 'threatened by technological changes' (1984, p. 19).

In their case-study of computing, Game and Pringle go one step further than both Kraft and Greenbaum, and seek to take account of those whose work is already 'deskilled'. They point to the fact that data entry clerks, for example, are so separated off from other computer-related occupations that very often they are not considered part of the same industry, an exclusion which can result in an obscuring of the role of gender relations in the structuring and organization of work in the computer industry (Game and Pringle, 1984, p. 82).[12]

The work of both Game and Pringle (1984) and that of Cockburn (1983; 1985), can be seen as bridging the theoretical gap in the gender and technology literature — between those that adopt a Marxist labour process framework, who continue to understand technology within a 'use-abuse' framework (as essentially neutral but misused under capitalism to deskill workers and thus increase managerial control over the labour process), and those that work, more explicitly, within a 'social constructionist' framework. Game and Pringle reject firmly the notion that technology is neutral. They argue that technology should be understood as the result of 'social processes . . . designed in the interests of particular social groups, and against the interests of others' (Game and Pringle, 1984, p. 17). Furthermore, and unlike the labour process approach, they understand gender as a social category, one that is 'fundamental to the way work is organized', just as work is 'central in the social construction of gender' (Game and Pringle, 1984, p. 14). Cockburn also consciously seeks to move away from deterministic approaches in her studies of gender and technology. Indeed, her central thesis in one of her major studies (Cockburn, 1985), is that people's relationship to technology and technical skills is part and parcel of what makes us men and women. Within this formulation, then, gender relations shape technology and technology, in turn, shapes gender relations.

Cockburn's research in particular, has been very influential in the development of feminist theories of gender and technology. She offers us serious critiques of both the equality and conventional labour process approaches in studies of gender and technological change. Her work makes clear that arguments for 'equal opportunities' contained within the equality approach are misguided because they fail to take into account the way in which male dominance of technology has been achieved by the active exclusion of women from areas of technological work. She brings the role of male workers to the fore, and explains the existing gender relations of technology in terms of both capitalism and hieriarchy.[13]

Despite, or rather because of, the important contributions made to the study of gender and technology (particularly in relation to work) by Cockburn, I want to explore further the implications of her arguments for change strategies. As we have already seen, Cockburn works within an explicitly 'social constructionist' framework, understanding both technology and gender as socially defined. She argues that technology and technical work have been defined as exclusively male activities in such a way that many tasks that women have traditionally performed — such as knitting — are not defined as technical despite involving a high degree of manual dexterity and computation (Cockburn, 1985). This observation seems to suggest that rather than arguing for women's inclusion in work currently defined as skilled and as technical, we should be arguing for a total re-evaluation of work so that many of women's traditional tasks are also recognized as skilled and technical and are given the appropriate remuneration. However, this is not the argument put forward most forcefully by Cockburn.

As McNeil has pointed out with reference to Cockburn's work, while her whole framework is one in which technological culture is examined and exposed for its gender dimension, she nevertheless appears, at times, to fall into the trap of 'taking technology at face value' (McNeil, 1987, p. 192). Indeed, McNeil argues that Cockburn's work has two underlying tendencies which should give us cause for concern: a certain belief in technological progress, on the one hand, and a certain acceptance of the assumption that women lack technological knowledge and skills, on the other. According to McNeil, these two tendencies reinforce each other so that Cockburn is able to argue:

> New technology is continually pushing up the minimum level of technical know-how that is saleable. Women have not yet got access to last year's knowledge, let alone next year's. (Cockburn, 1985, p. 162, cited in McNeil, 1987, p. 193)

There appears to be some overlap with the equality or 'women in technology' approach here, where technological developments are understood as giving rise to specific skill requirements that women must be able to meet if they are to enter technological jobs. Clearly, for Cockburn, women's need to access technological skills is not conceived in terms of individual advancement alone; she argues that knowledge of technology is 'personally empowering'. But does this not contradict her argument (discussed above) that technology and technical skills are socially defined in relation to gender? If skill is socially defined, would not redefinition, rather than access to areas of work currently defined as skilled, be the more appropriate strategy for change? Indeed, given what Cockburn and others (Hacker, 1989; Henwood, 1991b) have documented concerning women's experiences in technological work settings, there remain strong arguments for developing an alternative strategy.

One further issue is raised by Cockburn's understanding of the relationship between knowledge, technology and power. Cockburn has demonstrated clearly that dominating technological work confers upon men some material power over women, better paid jobs, for example, but there is a need to examine this relationship further. McNeil (1987) has suggested that the relationship between technology and power might not be that simple:

> Couldn't the obsessional knowledge of some working-class lads who are car buffs, or some of the avid readers of mechanics or computer magazines, be interpreted as evidence of impotence? (McNeil, 1987, p. 194)

McNeil argues for there to be a distinction made between 'representations' and 'realizations' of power. She suggests that while technology in our culture represents the promise of power, many 'will never realize the promise of technology' (p. 195). While I support McNeil in her attempt to demonstrate the importance of maintaining a distinction between the 'symbolic' and the 'material' significance of technology, and believe this distinction should be further developed and analyzed in feminist work, I want to ensure that we do not lose sight of the relationship between the two. For example, while the 'obsessional [technological] knowledge of some working-class lads' of which McNeil speaks may well represent their sense of lack of power in other areas of life, there does seem to be a positive relationship between their 'obsessional knowledge' and actual material power. After all, is it not men's dominance of these informal, leisure-based technological activities which gives them the confidence and experience to apply for technical training and employment? A struggle to overcome their relative lack of class power may, in this case, have led to a strengthening of their gender power.

The problem areas I have identified within Cockburn's work are common throughout the gender and technology literature and do, I believe, reflect the difficulty many of us have in thinking and writing about gender and technology completely outside of the dominant gendered discourses of technology. Technology is often taken at face value, as 'given', and the assumption that women lack technological skills is all too often accepted uncritically. Forms of both technological determinism and of essentialism persist. To use Linn's terminology, feminist analyses of technology have adopted both 'technology stereotypes' and 'gender stereotypes' (Linn, 1987). Linn has argued that just as gender stereotypes rest upon biology for legitimation, so technology stereotypes rest upon 'an equally partial view of production' (p. 151), focusing on hardware or artefacts. This tendency in much feminist work to date has serious implications for change. According to Linn, this tendency has led feminists to argue that technologies are designed and made by men, and (therefore) have negative consequences for women (p. 132). This particular construction has led to one of two strategies for change: first, a rejection of technology as 'masculine' or 'male'; second, a demand that women be given access to technological work so that they can redirect technology in a more 'feminine' direction. There are problems with both of these positions, not the least of which is their deterministic understandings regarding both gender and technology.

First, the simple rejection of technology position embraces a technological determinist view of the relationship between technology and society. Technology is understood as imbued with male values and thus is rejected. Linn has suggested that there exist strong elements of this view in feminist science fiction writing, where women reject male technologies, often in favour of a more 'woman-centered' lifestyle, in communion with nature.[14] By focusing in on the artefact and 'blaming the technology' in this way, this position would seem to rule out the possibility for changing the social relations of technology.

The second position is different; it does not reject technology but instead argues for more women in technology, because it is assumed that more women in technology will result in better technologies. Again, this position is evident in feminist science fiction writing,[15] but it is widespread in feminist non-fiction as well. It can be found in studies of women in engineering (Carter and Kirkup, 1990; van Oost and Everts, 1987; Kolmos, 1987), and is currently prevalent in studies of women and computing (see, for example, many of the papers in Lovegrove and Segal, 1991). The assumption here is that the entry of women into technological work will not only be good for women (as in the equality/women in technology approach), but will be good for technology. There is the suggestion here that, via the introduction of the 'feminine' into technological design, more appropriate technologies will emerge. Here, there exist elements of a social constructionist framework (the idea that feminine qualities can be introduced into technological design and thereby change technological priorities), together with essentialist understandings of gender (the equating of women with 'the feminine'). It seems to be impossible to avoid both technological determinism and essentialism at the same time.

Elements of this second position are currently emerging within the discourse surrounding many recent initiatives aimed at 'getting women in' to computing/IT. Articles in the computing press, as well as in the more academic journals, have begun to link the so-called 'changing requirements of computing', particularly the new emphasis on organizational and communications skills, to the demand for more women in the discipline (Hall and Lovegrove, 1988; Bruce and Adam, 1989). Here, we find elements of both technological determinism and essentialism: the technology 'requiring' certain specific skills which women are deemed to possess. Other commentators have gone further and suggested that changes within computer science itself (especially in the more prestigious areas of artificial intelligence (AI) and expert systems) may well offer women a unique opportunity to enter these fields (Huws, 1986; Perry and Greber, 1990). In particular, attention is drawn to the paradigm shift occurring in computer science in relation to theories of intelligence. It has been suggested that up until recently, the model of mental processes used in AI was a logical, linear and objective one but that this is gradually being replaced with a model of intelligence that includes 'flexibility, the ability to respond to unexpected situations in meaningful ways, and the ability to make intuitive leaps with insufficient information' (Perry and Greber, 1990, p. 94).

This shift, it is argued, has come about as a result of philosophical discussions about the difference between human beings and computers (Weizenbaum, 1976; Boden, 1977), but many commentators have suggested that it can be seen as offering, at the same time, a challenge to the historical association between masculinity and computers (Perry and Greber, 1990, p. 94). Does such a shift necessarily mean new opportunities for women? Again, it does so only in so far as we are able to equate women with these 'feminine' attributes now being seen as central in computing. While I would agree with Perry and Greber when they say that, 'Changes in the definitions of masculinity and femininity might arise from the reconceptualization of human beings in terms of the computer' (1990, p. 94), I would suggest that this will not necessarily lead to improvements in the position of women in relation to technology. It could equally well be that once these newly-discovered attributes of flexibility, intuition etc. are revalued and become sought-after skills in computing, men will be the first in line to demonstrate their competence in these fields.

Conclusions

It should, by now, be clear from the examples of feminist thinking about technology examined in this chapter that I believe existing frameworks for understanding gender and technology seem caught between the twin dangers of technological determinism and essentialism. It is all too easy, it seems, to accept technology at face value, to see technologies only as hardware or artefacts or a set of skills to be acquired. Similarly, it seems that many of us are all too ready to accept and work with essentialist understandings of gender when we believe that we can get women into technology on the 'feminine' bandwagon. However, when it comes to developing effective strategies for changing the gendered relations of technology, existing formulations do not seem to be serving us well. If we dislike the simple equality approach which focuses on 'getting women in', what are our alternatives? It seems that either we reject technology and with it the need for women to gain technological competence, or we argue that technology needs women because women are carriers of feminine attributes which technology currently requires. As I have suggested, both these positions seem to me to be dangerous and unlikely to effect lasting change in the gendered relations of technology, but where should we go from here?

In seeking new frameworks for the development of gender perspectives on IT, I believe that the newly emerging cultural analyses of technology can open up doors for us. I have already suggested that the contributions of McNeil and Linn represent a shift in feminist thinking about gender and technology. While still working within what we might term a social constructionist framework, their analyses represent a rather different approach from that of most feminist work to date, which has tended to understand technology in a material sense, as an artefact (albeit one that is shaped by the values of specific interest groups in society). McNeil and Linn, on the other hand, are arguing for greater concentration on the symbolic aspects of technology, and for an understanding of technology as culture. What is the significance of this shift?

Clearly, many studies of women and technology talk of 'the masculine culture of technology' and stress the ways in which boys and men dominate the design and use of technologies at home, in schools and in workplaces, how the language of technology reflects male priorities and interests, and how, therefore, women are effectively excluded from full participation in technological work. However, I see there being a fundamental difference between these studies, which examine technology and culture, where 'culture' refers to 'the environment' or social context, and is conceived of as separate from the technology itself, and those that examine technology as culture. In this latter formulation, technologies are themselves understood as 'cultural products', 'objects' or 'processes', which take on meaning only when experienced subjectively and where those meanings will vary according to the context in which particular technologies are encountered in everyday life. It is with these 'lived experiences of technology' that we need to be concerned.

Our theorizing of the gender and IT relationship should not be reduced to the simple 'man equals technology literate, woman equals technology illiterate' formulation. Technological meanings are not 'given'; they are made. Our task in trying to transform the gendered relations of technology should not be focused on gaining access to the knowledge *as it is* but with *creating that knowledge*. By this I mean we have to be involved at the level of definition, of making meanings and in creating technological culture. The option simply to reject technology is not open to us; Haraway has argued persuasively that technology is 'an aspect of our embodiment' and that we are all, therefore, 'cyborgs' (1985, p. 99), but how much have we really heard of cyborg women's experiences? We need more research that seeks to understand women's subjective experience of technology and take these as our starting point for our definitions of 'technology', 'technical work' and 'skill'. Similarly, we need more research on women's subjective experience of gender.

From my own research (Henwood, 1991b), which sought to examine women and men's subjective experiences of both technology and gender, I would suggest that any simple equation of men/masculinity/technological competence and women/femininity/technological incompetence, while exerting a powerful ideological force, does not correspond to many people's experience. My study suggests that while women do reject technological work because of its association with masculinity, many others are attracted to it precisely because of this association. Technological work, even when understood as masculine, does not have the same meanings for all women. In my study, 'masculine work' was understood as conferring status and power as well as 'manliness' upon its occupants, and women were found to have positioned themselves differently in relation to these meanings.

Femininity and so-called 'feminine work' were found to be similarly problematic for women in that study. Dominant discourses around 'equal opportunities' often assume (albeit implicitly) that women will want to move out of traditional women's areas of work and gain access to 'man's work', but in my study, a more contradictory picture was found. Women training for traditional 'women's work' were neither clearly for, nor clearly against traditional women's work. They showed ambivalence towards their chosen field (just as those in the non-traditional area had done). Here, the ambivalence was related to the fact that they perceived there to be both 'positive' and 'negative' aspects of femininity required by the job. A recognition of this ambivalence and the contradictory experiences arising from gendered work patterns can provide a starting point which can help us avoid simplistic analysis of, and solutions to, the 'problem' of women and technological work.

A suitable framework for analyzing gender and IT relationships then, is one which understands both technology and gender not as fixed and 'given', but as cultural processes which (like other cultural processes) are subject to negotiation, contestation and, ultimately, transformation. As such, they might be thought of as 'discourses'. Following Hudson (1984), I use the term 'discourse' to denote 'a snapshot of the matrix of ideas, terminology *and practices* surrounding an area of social life at a particular time' (Hudson, 1984, p. 34, emphasis added).

This emphasis on practices is important if we are to understand discourses as operating not solely at the symbolic level, but also at a more material level. A similar point is made by Walkerdine (1984) in her discussion of the discourses of masculinity and femininity where she argues that we need to be careful to distinguish between two levels of practice: 'Those practices which not only define correct femininity and masculinity but *which produce it by creating positions to occupy*' (Walkerdine, 1984, p. 182, emphasis added).

Thus, thinking about gender (and technology) as 'discourses' can be useful for it allows us to identify competing understandings and experiences while simultaneously leading us to the question of power. Discourses emerge out of struggles for power, both the power to 'define' but also, and more importantly, the power to create particular subject positions in line with these definitions. For example, the dominant discourse surrounding the 'women into technology' initiatives can be understood as redefining technological work in such a way as to create a position for women within technological work that allows them to be both technological and feminine. As I argued earlier however, so long as this discourse continues to define femininity as both synonymous with 'woman' and as unproblematic for women, it can only serve to further reinforce gender difference and divisions and impose impossible contradictions for the women who seek to identify with it. What we need is more research that seeks to explore the alternative, if more marginal, discourses of technology and to examine the factors responsible for the marginalization of the more radical and feminist discourses in favour of liberal ones that appear to reproduce and reinforce the *status quo*.

In relation to the specific issues being addressed in this book, concerned as it is with gender, IT and office systems design, what can be learnt from the arguments made here? Certainly, we should be sceptical about the value of a simple 'women in technology' approach which would argue for more women in systems design in the interests of equality or equal opportunities for women. In failing to address the source of the inequality, such approaches are only ever likely to benefit a few women who will continue to be seen as 'the exceptions that prove the rule' that women are not suited to technical work. Similarly, we should be wary of arguments that suggest that more women in systems design will, in and of itself, effect change in the design process itself and lead to new and more progressive priorities being adopted. What is important and constructive about focusing on the design process is precisely the way in which existing definitions of technology can be challenged and made available for negotiation and transformation, as much recent work reflected in the chapters of this book makes clear. Both technology and gender are constitutive of each other and we need to develop frameworks that are capable of analyzing the diverse ways in which the two interact to produce a range of different subjective experiences and practices. At its best, feminist research into the process of systems design can enable us to do this.

Acknowledgment

Thanks to Sally Wyatt who read and commented upon an earlier draft of this paper.

Notes

1 Some notable examples of such work can be found in the edited collections by Faulkner and Arnold, 1985; Kramarae, 1988; Hynes, 1989; and in Wajcman, 1991.

2 The survey was carried out by *Computer Economics* and was cited in Cowie, 1988.
3 The study surveyed 1,800 women professional members of the BCS and had 750 returns (a 40 per cent response rate). The findings of this study confirm what is already known about other professions where women are numerous but under-represented at the higher levels (see, for example, Skinner and Robinson, 1988, on social work and Al-Khalifa, 1988, on the teaching profession).
4 For example, many of the conference papers given, and discussions held, at the 'Women in Computing' (WIC) conferences over the last few years, suggest that WIC's overall strategy is essentially one of access and 'equal opportunities'. There has been a surprising lack of criticism about the WIT campaign objectives and strategies among this group. For a recent collection of WIC Conference papers, see Lovegrove and Segal, 1991.
5 There has been equal scepticism about 'WISE' initiatives, launched in 1984. WISE, or 'Women into Science and Engineering' is an Equal Opportunities Commission /Engineering Council campaign aimed at encouraging industry and education to set up education and training programmes to increase the number of girls and women entering science, technology and engineering education, training and employment. I have argued, elsewhere (Henwood, 1991a) that many WISE-type initiatives appear misguided (and are therefore unlikely to be fully effective) because they seek to encourage more women into science and technology without fully examining the reasons why there are not more women there in the first place.
6 In both her own *Pleasure, Power and Technology* (Hacker, 1989) and Smith and Turner's edited collection of Hacker's work, *Doing it the Hard Way* (1990), Hacker recounts how it was this experience at AT&T which moved her on, from a liberal to a socialist feminist framework in her research. (These two books also describe how she later found this framework unsatisfactory, too and moved on, via radical feminism, to a 'version of social anarchic feminism' (Hacker, 1989, p. 18).
7 It follows from this that a liberal feminist vision of the future would involve a situation where, through the provision of the same educational opportunities, every individual would be free to develop a full range of psychological qualities resulting in an androgynous society. Indeed, this future has been proposed by at least one feminist researcher concerned with encouraging women into engineering and technological work (Newton, 1987). For a detailed critique of this theoretical position, see Henwood, 1991b.
8 The 'Analytical Engine' was a 'programmable calculator', designed, though never built, by Charles Babbage in England in the 1830s and 1840s. It is now recognized as the forerunner to the modern digital computer developed 100 years later.
9 Electronic Numerical Integrator and Calculator, the first electronic computer, introduced in the USA in 1946.
10 Interestingly, Strober and Arnold's article notes that US Census classifications of occupations reflect this developing hierarchy in programming; in the 1960 and 1970 Census, all computer specialists, including programmers were classified as professionals but by 1980, computer scientists and systems analysts were in the professional category and computer programmers were classified as technical (Strober and Arnold, 1987).
11 This research provides empirical verification for the pioneering theoretical work on the relationship between sex and skill undertaken in the early 1980s by Phillips and Taylor (1980).
12 Indeed, they add that these data entry workers are often spatially segregated into separate companies which specialize in data preparation and may be located in a different town, or even country, from others in computer occupations. This point is verified by Posthuma's research on 'offshore office work' where big US companies were found to have 'decentralized' data entry tasks and made use of

suburban women's labour, as well as women's labour in 'off-shore' sites in the Caribbean (Posthuma, 1987).

13 As many commentators have now documented, this active exclusion of women from technological work has a long history in western capitalist societies. It is well-known that women took on technological jobs in both the First and Second World Wars, only to be actively excluded again after the wars (Summerfield, 1984; Braybon, 1981; Walby, 1986). While women's exclusion was the result of a 'deal' made between employers, trade unionists *and* the state, it was the power of the engineering unions to make their case — that the entry of women would lead to the 'dilution' of their skilled status — that determined those decisions. Both Game and Pringle's and Cockburn's work demonstrate that this pattern of exclusion and gender segregation continues today, ensuring that men retain control of high status technological work.

14 A good example of such an approach in feminist science fiction/utopia is Sally Gearhart's *The Wanderground* (Gearhart, 1979).

15 Linn (1987) suggests Sandi Hall's *The Godmothers* (Hall, 1982) and Susy Charnas' *Motherlines* (Charnas, 1979) are typical of this approach but there are many examples.

References

AL-KHALIFA, E. (1988) 'Pin Money Professionals? Women in Teaching', in COYLE, A. and SKINNER, J. (Eds) *Women and Work*, Basingstoke, Macmillan.

BEECH, C. (1990) *Women and WIT*, London, British Computer Society.

BODEN, M. (1977) *Artificial Intelligence and Natural Man*, New York, Basic Books.

BRAVERMAN, H. (1974) *Labor and Monopoly Capital*, New York, Monthly Review Press.

BRAYBON, G. (1981) *Women Workers in the First World War*, London, Croom Helm.

BRUCE, M. and ADAM, A. (1989) 'Expert Systems and Women's Lives: A Technology Assessment', *Futures*, October.

CARTER, R. and KIRKUP, G. (1990) *Women in Engineering: A Good Place to Be?* Basingstoke, Macmillan.

CHARNAS, S.M. (1979) *Motherlines*, Berkeley.

COCKBURN, C. (1983) 'Caught in the Wheels: The high cost of being a female cog in the male machinery of engineering', *Marxism Today*, **27**, pp. 16–20.

COCKBURN, C. (1985) *Machinery of Dominance: Women, Men and Technical Know-How*, London, Pluto Press.

COWIE, A. (1988) 'Screen Prejudice', *The Guardian*, 25 February.

CULLEY, L. (1986) *Gender Differences and Computing*, Department of Education, University of Loughborough.

DAIN, J. (1988) 'Getting Women into Computing', *University Computing*, **10**, pp. 154–7.

EISENSTEIN, Z. (1981) *The Radical Future of Liberal Feminism*, New York, Longman.

ENGINEERING INDUSTRY TRAINING BOARD (EITB) (1984) *Women in Engineering*, Occasional Paper No. 11, Watford, EITB.

FAULKNER, W. and ARNOLD, E. (Eds) (1985) *Smothered by Invention*, London, Pluto Press.

FOX-KELLER, E. (1986) 'How Gender Matters, or Why It's So Hard for Us to Count Past Two', in HARDING, J. (Ed.) *Perspectives on Gender and Science*, London, Falmer Press.

GAME, A. and PRINGLE, R. (1984) *Gender at Work*, London, Pluto Press.

GEARHART, S.M. (1979) *The Wanderground*, Watertown, MA, Persephone Press.

Greenbaum, J. (1979) *In the Name of Efficiency: Management Theory and Shop-Floor Practice in Data Processing Work*, Philadelphia, Temple University Press.

Griffiths, M. (1988) 'Strong Feelings about Computers', *Women's Studies International Forum*, **11**, 2, pp. 145–54.

Hacker, S. (1989) *Pleasure, Power and Technology*, London, Unwin Hyman.

Hall, S. (1982) *The Godmothers*, London, Women's Press.

Hall, W. and Lovegrove, G. (1988) 'Women and AI', *AI and Society*, **2**, 3, pp. 270–1.

Haraway, D. (1985) 'A Manifesto for Cyborgs: Science, Technology, and Socialist Feminism in the 1980s', *Socialist Review*, **80**, pp. 65–107.

Hartmann, H. (1979) 'Capitalism, Patriarchy and Job Segregation by Sex', in Eisenstein, Z. (Ed.) *Capitalism, Patriarchy and the Case for Socialist Feminism*, New York, Monthly Review Press.

Henwood, F. (1991a) 'Gender, Work and Equal Opportunities: Young Women and the Culture Of Software Engineering', in Lovegrove, G. and Segal, B. (Eds) *Women into Computing: Selected Papers 1988–1990*, Amsterdam, Springer-Verlag.

Henwood, F. (1991b) *Gender and Occupation: Discourses on Gender, Work and Equal Opportunities in a College of Technology*, Unpublished PhD thesis, University of Sussex.

Hoyles, C. (1988) *Girls and Computers*, London, Bedford Way Papers.

Hudson, B. (1984) 'Femininity and Adolescence' in McRobbie, A. and Nava, M. (Eds) *Gender and Generation*, Basingstoke, Macmillan.

Huws, U. (1986) 'The Effects of AI on Women's Lives', in Gill, K.S. (Ed.) *Artificial Intelligence for Society*, Chichester, Wiley.

Hynes, H.P. (Ed.) (1989) *Reconstructing Babylon*, London, Earthscan Publications.

Jagger, A. (1983) *Feminist Politics and Human Nature*, Totowa, NJ, Rowman and Littlefield.

Kolmos, A. (1987) *Gender and Knowledge*, in Engineering Education Paper contributed to the Fourth International Conference of Girls and Science and Technology, University of Michigan, July.

Kraft, P. (1979) 'The Industrialization of Computer Programming: From Programming to Software Production', in Zimbalist, A. (Ed.) *Case Studies in the Labor Process*, New York, Monthly Review Press.

Kraft, P. and Dubnoff, S. (1983) 'Software Workers Survey', *Computer World*, **XVII**, pp. 3–13.

Kramarae, C. (Ed.) (1988) *Technology and Women's Voices*, London, Routledge.

Linn, P. (1987) 'Gender Stereotypes, Technology Stereotypes', in McNeil, M. (Ed.) *Gender and Expertise*, London, Free Association Books.

Lovegrove, G. and Hall, W. (1987) 'Where have all the girls gone?', *University Computing*, **9**, pp. 207–10.

Lovegrove, G. and Segal, B. (Eds) (1991) *Women into Computing: Selected Papers 1988–1990*, Amsterdam, Springer-Verlag.

MacKenzie, D. and Wajcman, J. (Eds) (1985) *The Social Shaping of Technology*, Milton Keynes, Open University Press.

McNeil, M. (Ed.) (1987) *Gender and Expertise*, London, Free Association Books.

Newton, P. (1987) 'Who Becomes an Engineer: Social Psychological Antecedents of a Non-Traditional Career Choice', in Spencer, A. and Podmore, D. (Eds) *In a Man's World: Essays on Women in Male-Dominated Professions*, London, Tavistock.

Newton, P. (1991) 'Computing: An ideal occupation for women?', in Firth-Cozens, J. and West, M.A. (Eds) *Women at Work: Psychological and Organizational Perspectives*, Buckingham, Open University Press.

Perry, R. and Greber, L. (1990) 'Women and Computers: An Introduction', *Signs: Journal of Women in Culture and Society*, **16**, 1, pp. 74–101.

PHILLIPS, A. and TAYLOR, B. (1980) 'Sex and Skill: Notes Towards a Feminist Economics', *Feminist Review*, **6**, pp. 79–83.

POSTHUMA, A. (1987) *The Internationalisation of Clerical Work*, SPRU Occasional Paper No. 24, Sussex University, Science Policy Research Unit.

SHURKIN, J. (1985) *Engines of the Mind: From Abacus to Apple — The Men and Women Who Created the Computer*, New York, Washington Square Press.

SKINNER, J. and ROBINSON, C. (1988) 'Who Cares? Women at Work in the Social Services', in COYLE, A. and SKINNER, J. (Eds) *Women and Work*, Basingstoke, Macmillan.

SMITH, D.E. and TURNER, S.M. (Eds) (1990) *Doing It the Hard Way: Investigations of Gender and Technology: Sally L. Hacker*, London, Unwin Hyman.

STROBER, M.H. and ARNOLD, C.L. (1987) 'Integrated Circuits/Segregated Labour: Women in Computer-Related Occupations and High-Tech Industries' in NATIONAL RESEARCH COUNCIL (Ed.) *Computer Chips and Paper Clips: Technology and Women's Employment*, Washington DC, National Academy Press.

SUMMERFIELD, P. (1984) *Women Workers in the Second World War*, London, Croom Helm.

VAN OOST, E. and EVERTS, S. (1987) *Women's Interests in Engineering*, paper contributed to the Fourth International Conference on Girls and Science and Technology, University of Michigan, July.

WAJCMAN, J. (1991) *Feminism Confronts Technology*, London, Polity Press.

WALBY, S. (1986) *Patriarchy at Work*, London, Polity Press.

WALKERDINE, V. (1984) 'Some Day My Prince Will Come', in MCROBBIE, A. and NAVA, M. (Eds) *Gender and Generation*, Basingstoke, Macmillan.

WEIZENBAUM, J. (1976) *Computer Power and Human Reason*, San Francisco, Freeman.

WOMEN INTO INFORMATION TECHNOLOGY (1989) *WIT Newsletter*.

WOMEN INTO INFORMATION TECHNOLOGY (1990) 'The Women into Information Technology Campaign', paper presented to the Women into Computing Conference, University of East Anglia, July.

Section II

Gender Perspectives in Computer Systems Development

Chapter 3

Design of Information Systems: Things versus People

Susanne Bødker and Joan Greenbaum

Abstract

Information Technology is usually designed using traditional system development techniques and emphasizing conventional management objectives that focus on the information rather than on the people in a workplace. This chapter uses research from a gender perspective that highlights the ways that office systems can be designed with people in mind. It then applies the gender perspective to explain why Cooperative or Participatory Design can be used to enable system developers and office workers to work together to design applications that better support working practices.

The terms Information Technology and Office Systems are heavily balanced in the direction of focusing on things, instead of looking at people at work. Both the methods used to produce these systems and the actual computer systems in use reflect a strong bias towards looking at offices as if they were collections of systems that could somehow be solidified into a concrete object called information technology. In fact information technology's acronym, IT, speaks loudly to us about its focus. More than a century ago, Karl Marx described capitalism's fascination with quantifiable things. This fascination, which he called commodity fetishism, could be applied to the world of information technology, for the production of computer systems often looks like a process that develops, and sells information in packages of technology. The fault with this fetishism of information technology lies not only within the realm of capitalism as Marx described it, however. This bias towards the relationship between things, such as information, rather than the relationship among people, also grows out of the traditions of Western scientific thought that sweep social issues to the side of replicable, quantifiable and provable facts. Embedded within scientific work, as Evelyn Fox-Keller points out, are centuries of practices that push women's concerns with social interaction outside of the province of 'correct' scientific thinking (Fox-Keller, 1985). It is within this scientific tradition that the methods and practices of office system design were born.

Introduction

In this chapter we will be discussing the ways office systems are developed and in doing so, we will look a little at the background of this development process and then examine some of the ways that it is changing and we believe, can change. First we discuss what is wrong with IT from the perspective of the systems approach. Here we find that the standard system development methods support traditional management objectives of quantitative efficiency often at the expense of the people in the workplace. Next we look at how research from a gender perspective can give us ideas for how and why applications can be designed to better support the way people work. And lastly we take up the issue of cooperative or participatory design, suggesting ways that system developers and office workers can work together to develop applications that show respect for the workers involved and result in office applications that better fit working practice. One can argue that cooperative design is useful from a number of perspectives including both an ethical and a pragmatic perspective. For example, ethically we would say that cooperative design can build workplace democracy and help support the human rights of office workers. Pragmatically, we can say that including office worker's knowledge and experience in applications development results in technical applications that better fit working practices and are therefore more effective. Arguments about why or how a gender-based cooperative design strategy could be applied will, of course, vary depending on national and local traditions and problems. Here we present our analysis based on our experiences in this field.

As teachers, researchers and practitioners of system development we are caught up in the debate about how computer specialists and office workers (who system developers generally call the 'users') can better interact with each other. This debate is not new, although there is definitely national and cultural differences in the way it takes place. The way we see things, the increased knowledge, and to some extent influence of office workers, has reshaped the debate. Furthermore managerial strategies have changed over recent years. Both from within the US perspective of Participatory Design (Namioka and Schuler, 1992), and from the Scandinavian approaches to system development (Greenbaum and Kyng, 1991), the issue is no longer only about whether these workers should be involved in computer system design, but also, how their knowledge and experience can be put to good use. While debate in the systems field has shifted to enfold the interests of the people who will use the systems, the methods and suggestions are often stuck in the historical frame of computer science and its reliance on formal problem-solving. In this sense, many methodological suggestions are still glued to a traditional management perspective, namely that of control over the system development process. Furthermore they are, according to Christiane Floyd's (1987) framework, most often product oriented, rather than process oriented, emphasizing exactly the 'thing' focus of Information Technology.

While our field is called system development, with an emphasis on office or computer systems, we prefer to talk about computer applications, in order to keep our eyes fixed on how the technology is applied, rather than on the technology itself. Our research is empirical as well as theoretical and in this chapter we will draw on some of our experiences in developing and researching workplace applications. We are interested in getting a closer look at how people work together within workplaces and on how developers work with people who will be using the applications. In doing this we have tried to apply a gender perspective to the way work is done. We would like to stress that what we are describing here is not the mainstream of computer development and use either in Scandinavia or in the United States. Rather, our examples are openings, telling us that there is realistic hope for qualitative changes in how applications are developed and used.

We begin with the idea that computer applications are socially constructed, that is designed and used by people, and not driven by some technological need. In applying a gender perspective we do not choose to focus on differences between the men and women in the field, but rather on the way that all system developers are taught to apply methods that, we believe, perpetuate traditional and stereotypical dichotomies between women's concerns with social interaction and men's so called 'scientific concerns' with technology as a driving force. We believe that this is a false dichotomy that has driven a wedge between people and things, putting value on the latter and denigrating the former. Our use of a gender perspective tries to bring both values back into view with a hopeful eye toward developing computer applications that serve the needs of the people using them.

What's Wrong with 'IT'?

System development, and computer science, in general, is taught as a process of breaking problems down into manageable and controllable descriptions, a process which is the essential characterstic of what is called the systems approach. System developers learn to identify problems, solve them, and describe solutions. System development, in other words, is taught as a way of dividing and conquering the world, without getting emotionally involved in it. Yourdon, the author of many books on system development methods has this to say about the process:

> a design strategy that breaks large complex problems into smaller less complex problems and then decomposes each of these smaller problems into even smaller problems, until the original problem has been expressed as some combination of many small solvable problems. (1986, p. 61)

In this way, office systems, like other computer applications, grow up as abstract models of reality. The notion of the world as a 'system' is a way of reducing computer software and hardware as well as people to components, contributing to the overall system. These components are most often seen as describable in abstract terms. Another well-known system specialist describes the process this way:

> The inversion of viewpoints occasioned by Structured Analysis is that we now present the workings of a system as seen by the data, not as seen by the data processors. The advantage of this approach is that the data sees the big picture, while the various people and machines and organizations that work on the data see only a portion of what happens. (DeMarco, 1978, p. 49)

While many critics of office system design will argue that this focus on data flow is simply a mistake, the history of system development shows a long trail of similar 'mistakes', too frequent and too important to be ignored. As early as 1965, Robert Boguslaw attempted to warn the system development community, who he called the 'new utopians'.

> And so it is that the new utopians retain their aloofness from human and social problems presented by the fact or threat of machined systems and automation. They are concerned with neither souls nor stomachs. People problems are left to the after-the-fact efforts of social scientists. (1965, p. 3)

The history of modern scientific thought takes its origin in the period when science drew away from the so-called 'irrationality' of women's knowledge and towards attempts to control nature presumably through reason and scientific procedures (Fox-Keller, 1985). System thinking grew as the child of this scientific rationality (Ehn, 1990; Greenbaum and Kyng, 1991). It is within this scientifically oriented system development that the separation of people from things occurs, through incorporating formal sets of procedures for examining the things it identifies as data. Thus we believe that it is no accident or mere 'mistake' that office systems frequently don't fit the people who are to work with them. What is needed we believe, is a clear understanding of how these issues repeatedly come about, and what we, as office workers and system developers can do about it.

Applications Development — A Gender Perspective

> The most immediate issue for a feminist perspective on the natural sciences is the deeply rooted popular mythology that casts objectivity, reason and mind as male and subjectivity, feeling and nature as female. In this *division of emotional and intellectual labour*, women have been the guarantors and protectors of the personal, the emotional, the particular, whereas science — the province *par excellence* of the impersonal, the rational and the general — has been the preserve of men. (Fox-Keller, 1985, p. 2, emphasis added)

In *Reflections on Gender and Science* (1985), Evelyn Fox-Keller points out how societally-shaped notions of men and women strongly influence the way that science is done. It also lets us see that as science split emotional from intellectual labour, the scientific arena focused more and more on things, putting people into the background. In her work she shows us, both historically and in practice today, how these notions or myths place women and men on binary poles where their differences are emphasized in the questions asked by scientists and the methods used to answer those questions. These binary poles, or dichotomies, shape the man's world as supposedly rational, objective, and quantifiable, while the woman's sphere is painted as emotional, subjective, intuitive and qualitative. As with all severe dichotomies, proving or disproving their validity takes us away from the real issues we need to tackle. What is at stake is the fact that they shape our consciousness and therefore our behaviour.

Similar dichotomies can be seen in the field of system development. When we look at the procedures used to develop computer applications we generally see an overemphasis on seemingly objective criteria like hardware and software evaluation, with less attention paid to the so-called soft or subjective reactions of the people who use the applications (Greenbaum, 1990). As the quotations from Yourdon and DeMarco illustrate, computer system development is seen as a process of separating problems and cutting them up into manageable abstractions of reality. The development of office applications assumes that systems are so complex that they require scientific study, a form of examination that is somehow separate from daily life. This model building, which is supposedly rational and objective, leaves little room for those more people-oriented activities that women are socialized to be concerned with. In some ways the development of office systems can be seen as an embodiment of the separation of emotional from intellectual labour — a severing of the heart from the head. And superimposed on this frightening image is computer science's fascination, or fetishism with the head, and the abstract things it represents.

Our use of a gender perspective to the study of computer applications is not to bemoan the differences between men and women, for we feel that this unfortunately leads us back up the path where we find ourselves, once again, looking at 'women on a pedestal' or 'women as victims'. Nor is it to complain about how the fascination with things and models has pushed social issues to the side. Instead we will choose to reframe the way we go about looking at offices in order to begin mending the head and heart and going about the process of designing systems that better suit the people who use them. In the following we look at some of the ways where current research gives us pointers towards bringing people back into the picture.

Personal Involvement in Work

Here again we see Evelyn Fox-Keller as a guide. In her book about the biologist Barbara McClintock, she talks about McClintock's research methods that let one get a 'feeling for', or 'listening to', the subject of the research (Fox-Keller, 1983). Barbara McClintock marked her research with a strong belief that one should be 'letting the material speak to you'. Instead of descending on a research topic with a set of preconceived categories, McClintock chose to think of herself as being part of the material. For her, scientific work was living and being involved in the subject matter, or what she referred to as 'getting a feeling for the organism'. As system analysts we have tried to work in a similar way. We start from our personal involvement in our work and the way our work interrelates with others. For us applying a gender perspective means both getting involved in the workplace and looking closely at the people-to-people interactions taking place. As system analysts, we cannot, of course, ignore the technical dimensions involved in developing office applications, but we can, however, put these technical issues within the context of the social setting.

Complexity of Office Work

Elinor Wynn, a linguist took a similar approach in her study of clerical workers (Wynn, 1979, 1991). Noticing that many computer systems were designed to be 'idiot proof', as if clerical workers were somehow not very knowledgeable about their work, she undertook a study of clerical workers that highlighted the deeply complex and socially interactive nature of their conversation. Her work showed how seemingly 'simple' office conversation is rich in problem-solving and extremely important for getting work done. For system analysts Wynn's work illustrates how their reliance on scientific detachment may have pushed a focus on the information flow rather than on communication between people.

Wynn's work and that of many other researchers who study clerical work (US Office of Technology Assessment, 1985) help us look behind the false assumptions concerning women in the workplace. As other chapters in this book illustrate, office work is often seen by outside 'experts' as a series of trivial tasks done by women. In the eyes of many management consultants, women's work is almost invisible. The most stated argument used to support the introduction of office systems is that of efficiency, making office output faster and cheaper. When system developers begin with the assumption that this work is trivial, these so-called efficient systems often end up causing more and more stress and serious physical damage to the women involved. Recent studies indicate that, among other issues, the emphasis on faster keyboards and more printed output has resulted in painful injuries like carpal tunnel syndrome and, in some cases, levels of stress that compare with those found among air traffic controllers (Labor Institute, 1988). Certainly the pragmatic argument can be made that information technology that results in injuries and stress is not useful for increasing overall efficiency, nor does it result in better service. By recognizing the complexity and richness of office communication and interaction, system developers and office workers can help stir the essential ingredients of communication into the way office systems are developed, highlighting both the ethical need for more humane systems and the pragmatic concern for applications that actually support work practice.

Changing Situations and Shared Knowledge

Lucy Suchman's *Plans and Situated Actions* (1987) gives us another starting point for refocusing the design of office applications. Her work shows that human actions are not always guided by clearly defined plans, but are based on actions within specific situations. While office applications are often designed with the former in mind, Suchman's research reminds us that applications need to take account of how people react and exchange information depending on the situations they find themselves in. Thus a system designed as if information always flows from one department to another, may break down when clerical workers in one department find that they have to go around the system to get things done their own way.

Another guide-post in finding our way out of the gender stereotyped traps of traditional system development is the recognition that we need to find ways to better respect shared knowledge in addition to the more generally accepted authoritative knowledge. Lucy Suchman and Brigitte Jordan (1989) define authoritative knowledge as 'knowledge that is considered legitimate, consequential, official, worthy of discussion and useful for justifying actions. . .'. System developers and managers present tools like data flow diagrams and decision charts as authorative knowledge. Often when the authoritative knowledge of system developers clashes with the shared knowledge of office workers, the office workers' informal information gets buried within the 'legitimated' authoritative power structure. In recognizing the importance of shared knowledge and by acknowledging the significance of situation-based activity, system developers can better support people's working practices.

We believe that there is now more room to move application development in line with developing tools for how people work. This is obviously important from an ethical perspective, but now it is increasingly possible from a pragmatic perspective. In Europe and North America, for example, more and more office workers have some familiarity with computer applications. As they become familiar with applications they have also become more comfortable with voicing their concerns about systems that do not work. Increasingly managers are caught in a bind where they know that the data that comes from the Information Technology is not as good as the information the office workers have. There is room now for us to envision new ways of developing applications that go beyond the fascination with technical things. We feel that we are in a time where different paths are coming together giving both office workers and system developers the opportunity to set out in new directions.

Cooperative Design and the Integration of Emotional and Intellectual Labour

We do not believe that scientific rationality and authoritative knowledge have any better chance of withering away than Marx's predictions for the disappearance of the state under socialism. As computer scientists we know how well-grounded our discipline is in scientific and authoritative traditions. Yet we can clearly see that the ground is ready for system developers to begin to take seriously the issues that were uncovered in the preceding section. In fact, as we will discuss, these are no longer theoretical starting points, but rather issues that are being integrated into application design in both the United States and in Scandinavia. Three primary changes which have begun to occur include the need for system developers to:

- focus on the workplace and the actual practices of the people doing the work;
- involve office workers at all levels in articulating their needs and expressing their concerns for what kinds of computer support they may need;
- develop new methods that help system developers and office workers actively support ongoing social processes.

These issues are a few of the wide range of approaches coming together under the rubric of Cooperative or Participatory Design (see Greenbaum and Kyng, 1991). While Cooperative Design is not the mainstream of practice in the system development field it does present ideas and practices that are being increasingly recognized. The cooperative approach means that computer application design activities begin and end with the people in the workplace. It includes training system developers to look at office workers as competent practitioners rather than invisible entities. Thus a payroll system that formerly had been designed to issue cheque and documents automatically 'without human error', now could be designed to allow payroll clerks to look up information and answer inquiries about problems like lost cheque or incorrect tax codes. The resulting increase in quality of service could be viewed with equal, and perhaps greater importance, than the solitary focus on greater speed and increased output.

The importance of involving office workers in the design process has begun to take on added significance (see Friedman, 1989). Traditionally, users of computer systems were seen as the managers ordering the system. Increasingly, however, as the term 'end-user' has crept into computer jargon it is apparent that both managers and system developers have begun to realize that the 'end-user', or the person who actually does the work, needs to be included in the design of information applications (Grudin, 1991). Sometimes this means that secretaries and clerks are only asked to describe their work tasks, but from the perspective of Cooperative Design it increasingly means that these workers need to get involved and stay involved with the system developers as the new application is emerging.

Which brings us to the third point, that of finding new methods that help office workers articulate their needs and enable system developers to listen to them. It is in this area that we have had the most experience, and it is precisely this area that has been propelled along by help from a gender perspective. Here are a few examples of the types of changes we have seen taking place.

In a study of system development work at a large bank in Denmark (Bødker and Greenbaum, 1988), we conducted a series of workshops with system analysts to help them find out what improvements they wanted in their work. Among other requests was their frustration with only seeing a part of the development process, and being cut off from an understanding or overview of the project. They argued that if they had a clearer view of the whole project they would be in a better position to understand bank operations and therefore sort out some of the problems that the bank workers had with previous system design. As one woman pointed out:

> It is embarrassing, when my friends ask me what I do, I tell them that I make systems for the bank. When they ask me how that affects them as customers in the bank, I must say I don't know! I don't know how the programs I make will work in the bank office. What I do know is how my modules interact with those programmed by others in the group.

In addition they urged that since development work was project or group oriented they needed to have a 'feeling of belonging' to the group. This feeling of belonging, like their frustration with piecemeal system development is not addressed in traditional systems literature, but is certainly a fundamental aspect of feminist research, for it recognizes the need for ongoing social processes that support the way people actually work.

In another project we saw that office workers who were given the chance to talk about their ideas concerning desk-top computers, gained a lot from the process and were able to express their needs for the type of hardware and software they wanted their organization to use (Greenbaum and Madsen, 1992). The project involved a dozen US office workers in a wide variety of titles from secretaries and editorial assistants to accounting clerks and editors. They participated in a series of workshops modelled after a technique called The Future Workshop

(Jungk and Mullert, 1987), which encourages participants to speak freely among themselves and find constructive alternatives to their problems. The participants in the US office, like many others that have been involved in workshop environments found that their problems with the organization's current computer applications were not their own individual problems, but rather shared concerns that they could begin to do something about. Their use of this shared knowledge enabled them to find ways to confront what had previously seemed to them like a wall of authoritative knowledge from the systems department (Greenbaum, 1991). They were able to listen to each other and get, in McClintock's words 'a feeling for the organism'. From the standpoint of computer application development, they were able to better express their needs and concerns about the type of desk-top hardware and software they needed to get their jobs done.

In one case, in a Danish public office, we have seen a move from centralized system development to decentralized development where each branch office is provided with fourth generation tools for their own use. These so-called fourth generation tools are, among other things, computer languages that are designed for use by non-computer specialists. In this decentralization process a small number of workers from the branch office were trained as 'computer instructors'. These instructors then act as both local system developers and continue to take part in the everyday business of the office (see Bødker *et al.*, 1991). As with many systems this change was done to save money. At the same time, we see in the move an acknowledgment of the need for an anchoring of system development in the daily work of the office workers. Additionally, the office workers involved got a chance to learn new computer skills and to participate in designing the applications that they use. This case is still underway. It will be interesting to see if the office workers find it useful to act as computer instructors, and to what extent they feel overburdened by additional work. Much work remains to be done to find suitable ways of working with system development in decentralized work environments. (A research project focusing on these issues is outlined in Bødker *et al.*, 1991).

To support the increasing involvement of office workers in computer application development further, system analysts have begun to use prototypes or trial systems to give workers a better chance to see and experience what the system will actually do. In one case, for example, with dental assistants, the system analysts developed computer prototypes that showed pictures of teeth so that the assistants could refer to the visual charts on the computer screen when they were working with patients (Bødker and Grønbæk, 1991a, b and c). As it turned out in this case, the system analysts had developed a 'picture' of the mouth that was upside down from the way the dental assistants viewed patients' mouths. The problem was quickly solved since the prototype was merely a trial system and not the result of some long drawn-out traditional development project. Using prototypes lets workers get involved in very concrete aspects of application development, rather than relying on the abstract and often detached knowledge of system analysts. This use of prototypes is gaining more acceptance in system development literature and practice. We would like to see more office workers get involved in testing prototypes, because we feel that this is an important issue for demanding more rights for office workers and for bringing about more people-oriented office applications. This may sound like a like a lot, particularly in rough economic times, but from our experience a common denominator among office workers and systems developers has been the desire to produce quality products and services.

While we could continue to go on offering examples of system analysts and office workers trying new ways of working together and sharing knowledge, we will end here and simply summarize our analysis of what is happening. Clearly the recognition of the need for office workers to get involved in the design and use of office applications is gaining popularity in the computer field. Over the last few years there have been a number of conferences, journal articles and trade publications addressing these issues (see for example Namioka and Schuler, 1992; CSCW'90 Proceedings; and recent issues of *Communications of ACM*). This is a good start. These changes are taking place because office workers are getting more knowledgeable about office applications and demanding to get involved. They are also occurring as system analysts and managers realize that the IT that they have developed often does not fit the needs of the workplace. At this point in time, moving toward Cooperative Design is more of a political process than a technical one. Changes in work organization and relative power relations of office workers vary from country to country. We hope that this chapter begins to make a start toward ways that office workers and system analysts can envision more people-oriented computer applications and new ways of working together.

References

BØDKER, S. and GREENBAUM, J. (1988) 'A Non-Trivial Pursuit, cooperation in systems development', in KAASBØLL, J. (Ed.) *Information Systems Proceedings* (IRIS), Rørus, Norway.

BØDKER, S. and GRØNBÆK, K. (1991a) 'Cooperative Prototyping Studies — Users and Designers Envision a Dental Case Record System', in BOWERS, J. and BENFORD, S. (Eds) *Studies in Computer Supported Cooperative Work: Theory, Practice and Design*, Amsterdam, Elsevier Science Publishers/North Holland, pp. 315–32.

BØDKER, S. and GRØNBÆK, K. (1991b) 'Cooperative Prototyping: Users and Designers in Mutual Activity', *International Journal of Man-Machine Studies*, **34**, Special Issue on CSCW, pp. 453–78.

BØDKER, S. and GRØNBÆK, K. (1991c) 'Design in Action: From Prototyping by Demonstration to Cooperative Prototyping', in GREENBAUM, J. and KYNG, M. (Eds) *Design at Work: Cooperative Design of Computer Systems*. Hillsdale, NJ, Lawrence Erlbaum Associates, pp. 197–218.

BØDKER, S. *et al.* (1991) 'Computers in Context — Report from the AT-project in Progress', *Proceedings of the NES/SAM Conference*, Ebeltoft, Denmark.

BOGUSLAW, R. (1965) *The New Utopians — A Study of System Design and Social Change*, Englewood Cliffs, NJ, Prentice-Hall.

CSCW'90 (1990) 'Proceedings of the Conference on Computer Supported Cooperative Work', Los Angeles, California, October 7–10, ACM Press.

DEMARCO, T. (1978) *Structured Analysis and System Specification*, Englewood Cliffs, NJ, Prentice-Hall.

EHN, P. (1990) *Work-Oriented Design of Computer Artifacts*, Hillsdale, NJ, Lawrence Erlbaum Associates.

FLOYD, C. (1987) 'Outline of a Paradigm Change in Software Engineering', in BJERKNES, G. *et al.* (Eds) *Computers and Democracy — A Scandinavian Challenge*, London, Avebury.

FOX-KELLER, E. (1983) *A Feeling for the Organism: The Life and Work of Barbara McClintock*, San Francisco, W.H. Freeman.

FOX-KELLER, E. (1985) *Reflections on Gender and Science*, New Haven, CT, Yale University Press.

FRIEDMAN, A. (1989) *Computer Systems Development, History, Organization and Implementation*, London, Wiley.

GREENBAUM, J. (1990) 'The head and the heart', *Computers and Society*, **20**, 2, June, New York, ACM.

GREENBAUM, J. (1991) '*The head and the heart revisited: Towards Participatory Design*', in ERIKSSON, I.V., KITCHENHAM, B.A. and TIJDENS, K. (Eds) *Women, Work and Computerization: Understanding and Overcomming Bias in Work and Education*, Amsterdam, North-Holland.

GREENBAUM, J. and KYNG, M. (Eds) (1991) *Design at Work: Cooperative Design of Computer Systems*, Hillsdale, NJ, Lawrence Erlbaum Associates.

GREENBAUM, J. and MADSEN, K.H. (1992) 'Small Changes' in NAMIOKA, A. and SCHULER, D. (Eds) *Participatory Design*, Hillsdale, NJ, Erlbaum Associates.

GRUDIN, J. (1991) 'Interactive Systems — Bridging the gaps between developers and users', *Computer*, April, Los Alamitos, CA, IEEE.

JUNGK, R. and MULLERT, N. (1987) *Future Workshops: How to Create Desirable Futures*, London, Institute for Social Inventions.

LABOR INSTITUTE (1988) *A Price for Every Progress*, (video), New York Institute for Labor Education and Research.

NAMIOKA, A. and SCHULER, D. (Eds) (1992) *Participatory Design*, Hillsdale, NJ, Lawrence Erlbaum Associates.

SUCHMAN, L. (1987) *Plans and Situated Actions: The Problem of Human-Machine Communication*, Cambridge, Cambridge University Press.

SUCHMAN, L. and JORDAN, B. (1989) 'Computerization and Women's Knowledge' in TIJDENS, K. *et al.* (Eds) *Women, Work and Computerization*, Amsterdam, North Holland.

US OFFICE OF TECHNOLOGY ASSESSMENT (1985) *Report on Office Automation*, Washington US Congress.

WYNN, E. (1979) *Office Conversation as an Information Medium*, (dissertation), University of California, Berkeley.

WYNN, E. (1991) 'Taking Practice Seriously', in GREENBAUM, J. and KYNG, M. (Eds) *Design at Work*, Hillsdale, NJ, Erlbaum Associates.

YOURDON, E. (1986) *Managing the Structured Techniques*, New York, Yourdon Press.

Chapter 4

A Separate Reality: Science, Technology and Masculinity

Fergus Murray

Introduction

This chapter explores relationships between masculinity and technology. In particular, I address these relationships with reference to various parts of the Information Technology field.

The exploration I have undertaken can hardly be described as a voyage of discovery. A more apt analogy might be a stumble around the dimly lit broom cupboards of my own and other people's work and experience.[1] In the process I feel I've disturbed a lot of dust. Emerging back in the light I find I've gathered a handful of jigsaw puzzle pieces. The pieces won't fit together and I wonder if they are from different jigsaw puzzles. The different pieces are suggestive of different themes. Perhaps they are equally important, perhaps the broom cupboards were built by post-modernists, perhaps there is no one perspective that can hope to explain the multi-faceted relationship of technology, science and masculinity. This emerging realization has been irritating and confusing; I am not immune to the desire to make a coherent picture, to fix it in time, and distil complexity and uncertainty into a solid and totalizing theoretical framework.

This chapter then is a presentation of different pieces of a jigsaw puzzle. At this point I don't know if they are from the same puzzle. Anyway, here is a brief introduction to the pieces.

Here is a piece that comes from a research area that studies masculinity; it has a lot of interesting things to say about masculinity but precious little to say about science or technology. It suggests that masculinity is struggled for, is social rather than biological, and is itself heterogeneous. Masculinity changes over space and time. Masculinity is context specific, but wherever it struts around it does so in a difficult relational dance with femininity.

Another of the jigsaw pieces on my desk suggests stark contrasts between men and women and science and technology. In essence it asserts that men need the heroic voyage of science and technology because they cannot bear children. This piece is provocative but without the rest of the puzzle it looks improbable. It appeals to a kind of ahistoric psychoanalytic theory of womb envy.

A third piece concerns the psychological make-up and biographical background of engineers. Perhaps the womb envy theory is too broad. After all, not all men become scientists and technologists even if all men can draw on a cultural identification of science and technology with masculinity. But if we are to stay with a psychoanalytic approach we need a key to understanding why some men get into this science and technology stuff in a big, consuming way and others don't. So, maybe part of the answer lies in childhood experiences and, in particular, a flight into the controllable world of technology and scientific certainty as compensation for early experiences of loss of control, of the overpowering chaos of the social. This doesn't explain why boys get to rush into science while girls get family rearing or narrow ranges of low paid 'women's work'. But maybe there is something to be gained from asking as complementary questions to those already raised: *which* men go into science and technology?

Two other jigsaw pieces look at what men say about their attachment to making technological artifacts; the third looks at the way masculinity helps shape the process and organization of making technology. These pieces raise a different set of questions: What is it that men get out of making technology above and beyond well-paid, high status jobs? Does the field or context of technical activity influence the emergence of different masculinities? And, are some technologies and technologists more masculine than others and, if so, why?

These are the pieces I have picked up and puzzled over in my wander through the fields of masculinity studies and technology studies. At present I am unwilling either to discard any of the pieces or to bash them together with a wooden mallet in the hope that they form a coherent picture. Instead, I want to look in more detail at each of them in turn.

Masculinity

This section of the chapter looks at the concept of masculinity. What is it? Where does it come from? Is it fixed or mobile, unitary or plural, stuck, stubborn and wholly negative or open to the possibility of change?

In its broadest sense masculinity is the way men behave; it is the way men think and feel about themselves. Far from being a natural or biological category, masculinity is a socially constructed way of seeing and being. As such it can and does change over both time and space. For example, concepts of manhood in medieval and contemporary times have changed considerably, and they also differ between cultures and ethnic groupings. Furthermore, there are different conceptions of masculinity within the same society. These to an extent will depend on factors such as class, race, place and sexual orientation.

If masculinities are plural and socially constructed it follows that masculinity is not automatically conferred upon men. It doesn't simply come with a penis at birth. And because it is relational and heterogeneous there is always the possibility of not having enough of it (and thus slipping into the feminine) or having the wrong sort and being at risk of attack from other hegemonic masculinities (Connell, 1987). Masculinity is something that men struggle to achieve and maintain in highly competitive circumstances. Vic Seidler says, 'Gender is not something we [as men] can be relaxed and easy about. It is something we have to constantly prove and assert' (1989, p. 151).

Masculinity is a relational concept. It only makes sense, indeed it can only be defined, in relation to femininity. Masculinity and femininity are locked in a dance where their respective positioning constrains the space within which the

Figure 4.1: The masculine and feminine single boundary model

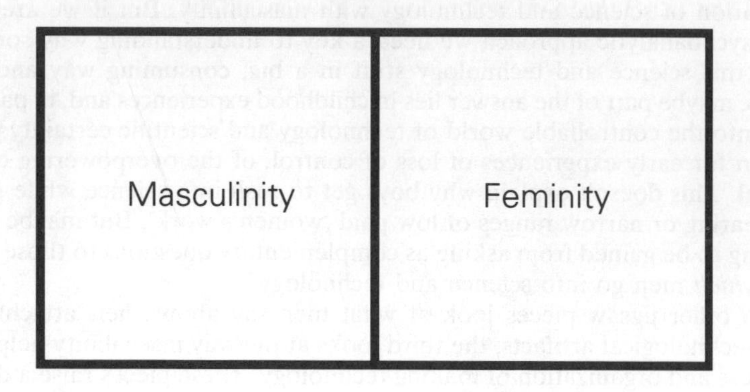

Figure 4.2: The masculine and feminine gradation boundary model

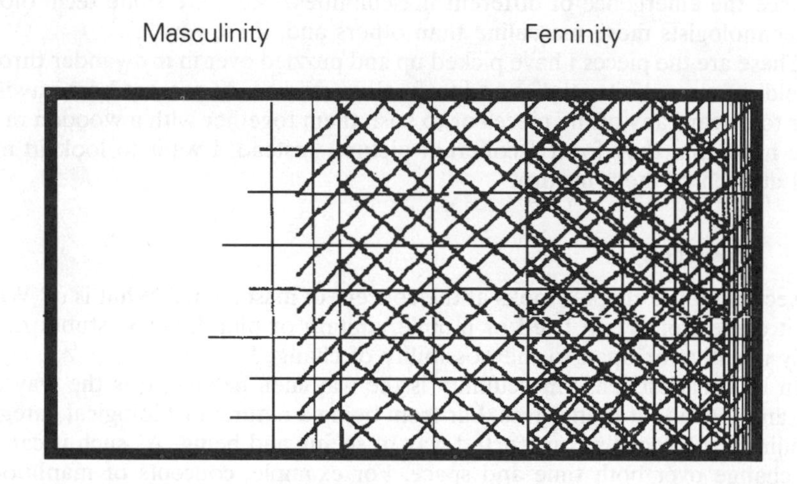

other can define itself. However, the dance involves more than an uneasy partnership of a single masculine and feminine identity. It is more like a crowded club where different masculinities and femininities jostle and fight among and between themselves.

If we accept that there is masculinity and femininity and that they are relational, it follows that there is somewhere a boundary, or boundaries, between masculinity and femininity. Such a boundary would mark where masculinity stopped and femininity began. This could be represented graphically as in Figure 4.1. This representation suggests a single, hard and fast boundary between one masculinity and one femininity. A different way of representing this relationship is shown in Figure 4.2. This suggests that there are subtle gradations between the core domains of masculinity and femininity. Boundaries are multiple and often ambiguous. Different masculinities may feel more or less tolerant of this ambiguity. Some may determine to maintain particular boundaries, others may be indifferent to their trangression.

In this chapter I want to suggest that masculinity tries to claim as a core domain exclusive to itself the practice and culture of science and technology. This is closely associated with masculinity's claims to rationality (Seidler, 1989). It is also linked to masculinity's alienation from the body and emotion (Corneau, 1991).

In arguing that technology is a core domain of a socially constructed masculinity I want to suggest that it plays an important role as a boundary marker; what is perceived to be technological is perceived to be masculine. That is, masculinity claims for itself an exclusive control of the technological and when masculinity fails to control or loses control of technological practices those practices then lose their status as technological practices.

I am not suggesting that masculinity's claim to represent technology is automatically achieved. Indeed, it is highly contested. But in general terms, and despite the challenges to it, there is still a strong relationship between masculinity and technology. This takes on the appearance of a natural truth and forms an important justification of men's control and direction of technological practices.

In this chapter I want to look at various aspects of this relationship. In part, I am concerned to examine the way in which masculine and technological practices intermesh and reinforce each other. But, perhaps more importantly, I am interested to explore the subjective and symbolic significance of technology as artefact, practice and culture for masculinity. In this, I want to ask the question: what does technology symbolize and mean for masculinity and for men?

In the remainder of this chapter I want to explore the linked areas of the symbolic and subjective significance of technology for masculinity and the control of technology practices by men. Initially, then, I look at work on technology that either explicitly or implicitly explores the subjective significance of technology for masculinity. I then go on to look at the interweaving of masculine and technology practices in the area of software development.

Fathering the Unthinkable

In the early 1980s Brian Easlea wrote a polemical book called *Fathering the Unthinkable: Masculinity, Scientists and the Nuclear Arms Race*. In it Easlea argued that men's infatuation with science and technology was largely a compensatory mechanism for their inability to bear children and their sexual vulnerability. As a defence against ridicule and loss of self-esteem men make things rather than babies: big, sophisticated, controllable and controlling technological things. Easlea says,

> Men just cannot give birth to children . . . and their masculinity is always vulnerable to female ridicule, particularly if and when sexual penetration is attempted. Alas for the psychologically beleaguered male. From preliterate to highly literate society, the male dilemma remains basically the same. (1983, p. 18)

In essence, Easlea argues that the male dilemma — the inability to bear children — is timeless and universal. As such this psychoanalytical approach relies on an implicit theory of womb envy that appears impervious to history, context or change. It is a constant although the results of its drives change over time: yesterday's medicine man is today's nuclear physicist.

Easlea argues that scientific discourse develops a vivid and aggressive imagery of invasion. This involves the occupation and dissection of a passive and mysterious female 'nature'. Here science developed as a distinctively masculine activity where the 'deeper the mental penetration into female nature the greater the mental virility the man of science is able to claim' (1983, p. 171). This gives rise to a hierarchy of potency and status within the sciences where the most penetrative and dissecting activities such as particle physics stand above the 'softer' systemic approaches such as biology and ecology.

Seen from this perspective science is a cold, dry, hard, aggressive activity that glories in its own penetrative abilities in the pursuit of a complete 'mastery' over nature. Symbolically, the pursuit of science appears to be an attempt to create a controllable and knowable world. This world has clear, rigid boundaries. In it different parts of the system ('nature') are linked by explicit and logical rules.

Science and masculinity in the modern era, then, appear to develop together. This involves not only the dissection of a socially constructed female nature but also the self-mutilation of the potentiality of a different kind of masculinity. For men's subjugation of a feminine nature proceeds apace with a need to 'subjugate and conquer the feminine within themselves' where this includes the need to relate, to enter into dialogue, receptivity (listening/empathy), and the validation of and involvement in 'simple domestic concerns' (Easlea, 1983, pp. 146, 37).

Easlea's work in *Fathering the Unthinkable* opens up important ground for investigation but leaves many significant stones unturned. For example, the shift from the study of Francis Bacon's ideas and Mary Shelley's allegorical writing to nuclear physics is, to say the least, sudden. This snapshot approach to history is provocative but does not examine the more complex and contested social processes that have given rise to shifting relations between masculinity and science (see Jordanova, 1987). For Easlea it appears as if the relationship of masculinity and science and technology was established and fixed with the emergence of Baconian science in the seventeenth century.

Historically grounded accounts of the development of science have pointed to the incursion of women into this privileged masculine realm. However, Rossiter (1982) concludes that in the period 1880–1910 women's position within science was constrained in two ways: they were either limited to holding subservient positions as assistants and educators or confined to practise science in 'women's' fields such as home economics or cosmetic chemistry.

Harding (1986) also examines the subordination of women in the sciences in this and the post-war period. She argues that the cultural stereotype of science as tough, rigorous, rational, impersonal, competitive and unemotional has continued to be 'inextricably intertwined with issues of men's gender identities' in a mutually reinforcing manner (1986, p. 63). Science and technology not only seems to vest masculinity with a particular potency in, and claim on, the world; it also renders masculinity and science particularly vulnerable to feminine 'dilution'. Harding says,

> We should expect that in science more than any other occupation (except, perhaps, making war) it will take the presence of only a very few

> women to raise in men's minds the threat of feminization and thus of challenges to their own gender identity. (1986, p. 63)

Harding appears to be saying that science offers masculinity 'something' particularly important at a cultural and perhaps subjective level. So important is this 'something' that even the threat of feminization provokes stauncher resistance than in other areas of masculine practice. This 'something' that science and technology offers masculinity is more than economics and the control of work and professions. It is about a deeper identification where scientific method and culture seems to echo and legitimize some essential masculine beliefs and practices. The following sections of the chapter look in more detail at what this 'something' might be.

The Soul of the New Machine or a Separate Reality

This section of the chapter examines an ethnography (Kidder, 1982) of a computer development project and recent research on engineers (Hacker, 1990). In it I raise two questions: why do men become engineers and IT specialists; and what do they get out of their profession at a subjective level? In particular, I am interested in the connection between what we might call the psychic security that derives from these specialisms and the way in which it links with prevailing modes of masculinity. Sally Hacker has explicitly addressed this concern in her work on engineering students. Tracy Kidder has not but the account he provides of the development of a 32-bit mini-computer provides some interesting, if inconclusive, pointers in this area.

Sally Hacker set out to investigate the social formation of engineers through a comparative study of engineering students at the Massachusetts Institute of Technology (MIT). The aspect of this work I find most interesting is Hacker's exploration of the relationship between the childhood experiences of the engineering students and their decision to enter the profession. For example, she found that the engineering students, 'painfully recalled children's bodies that would not do what they should' (Hacker, 1990, p. 115). In addition, they experienced difficulty remembering the sensual pleasures of childhood in contradistinction to the humanities students she interviewed. From these and other findings she concludes that,

> The men who chose engineering had early life experiences that emphasised aloneness, that allowed them greater distance from intimacy or the pleasures and dangers of 'mixing it up' with other people. Many became fascinated by things, and how they worked. These experiences heightened the value placed on abstractions and the control of nature. (Hacker, 1990, p. 124)

Hacker here puts forward a very interesting thesis: that there may be early childhood experiences that take place within particular familial and societal contexts that predispose some individuals to become engineers. Further, these individuals grow up and seek psychic security in the world of things almost as a compensation for early disappointment or trauma. In support of this thesis Hacker quotes an engineering student who said,

> My mother was a hysterical woman . . . . I think somewhere along the line I felt the need for things you could trust . . . . That was the attraction of mathematics. (1990, p. 117)

Notwithstanding our reservations on the relationship between 'hysterical mothers' and engineering careers Hacker raises an interesting issue that is worthy of further research.

Tracy Kidder's ethnography of an almost exclusively male computer development team touches anecdotally on this area; there is a link in the early life experiences of some of his subjects and their drift into computing. In particular, a number of the development team appear to have experienced themselves as failures either in terms of their sporting or academic achievements. Others found a reassuring solidity in the world of things. For example, one of the senior team members was a small, pale, weak child who felt himself to be at the 'bottom of the pile'. When he worked out how to take apart a telephone at an early age he said, 'This was a fantastic high, something I could get absorbed in and forget that I had these other social problems' (Kidder, 1982, pp. 87–88).

The project manager, branded an 'underachiever' at college, found a kind of security in the world of things. He said, 'There's some notion of control . . . that you can derive in a world full of confusion if you at least understand how things get put together' (Kidder, 1982, p. 158).

Kidder's ethnography only touches on these early experiences of the minicomputer engineers in question. It provides more material on the attraction of being a computer engineer. For despite the long, gruelling hours and poor working conditions the computer engineers in Kidder's study are strongly attached to their work. As one says, they were breathing life into a new machine and by so doing making something bigger than themselves. They were also at the very cutting edge of technological developments and in some senses were making history. This sense of going where no man has gone before provides considerable satisfaction to many engineers. One of the engineers Cynthia Cockburn interviewed said,

> You collectively are at the forefront of whatever it is you're doing . . . . It doesn't have to be anything really [!] wonderful. The fact that you are at the limit of your company's experience: That's a marvellous feeling. And if it is the case, as it was with me, that you are involved in something which, without being mealy-mouthed about it, is doing humanity a bit of good. The scanner was such a tremendous breakthrough for the medical world. Everybody was so excited, swept up in it. It was terrific. (Cockburn, 1985, p. 175)

However, such is the speed of technological change that this sense of being at the leading edge may be short-lived. This was the case for the project manager in Kidder's story of computer development. He stressed the tentative character of his 'marvellous feelings' when confronted with machines he had designed earlier in his career. These now appeared 'clumsy' and their makers 'dumb'. In a revealing quotation he said, 'You spend all this time designing one machine and it's only a hot box for two years, and it has all the useful life of a washing machine' (Kidder, 1982, p. 162).

The thrill of designing machines derives from the novelty, from the knowledge that no one has done this before, and that it will command the status of a 'hot box', the top kid on the block, for a couple of years. The project manager does not seem concerned with the utility of his creations. Instead, he's horrified by the way in which his cherished babies so quickly become ugly and lose their specialness, in which they become mundane and domesticated.

Working on the latest technologies invests lives with a particular significance. The flight test pilots in Tom Wolfe's (1980) study of US astronauts faced huge risks to push at the 'edge of the envelope' of jet flight. The thrill of being at this edge is closely bound up with the status bestowed on test pilots who have the 'right stuff', but it is often a precarious and short-lived thrill. Wolfe's descriptions of pilots burnt beyond recognition and the perils of ejecting from a jet fighter out of control are horrendous. And in the 1960s an ordinary US Navy pilot, let alone a test pilot, faced a 23 per cent chance of dying in an accident.

Test pilots also flew under conditions of intense competition. In their elite and mysterious fraternity of flying, booze and fast driving, pilots vied to climb the increasingly dangerous steps of a ziggurat of the right stuff. One slip and you were either incinerated or flying military transports — the collective flight test pilot's equivalent of a living death.

Having the right stuff is an intense source of pride for the test pilots. Yet it is never referred to openly. It's a hidden code, an unwritten text of righteous manhood. Wolfe describes it like this:

> The idea here (in the all-enclosing fraternity) seemed to be that a man should have the ability to go up in a hurtling piece of machinery and put his hide on the line and then have the moxie, the reflexes, the experience, the coolness, to pull it back in the last yawning moment — and then to go again *the next day*, and every next day, even if the series should prove infinite — and, ultimately, in its best expression, do so in a cause that means something to thousands, to a people, a nation, to humanity, to God. (1980, p. 29; emphasis in the original)

The flight test pilots lived in a hermetically sealed world of brotherhood and death. In the world of machine making this detachment is perhaps not as extreme. But there are moments in Kidder's story where 'ordinary life' is almost completely excluded. The one woman engineer on the mini-computer project comments, 'You can end up staying all night. You can forget to go home and eat dinner' (p. 61); a Microcoder on the project working on a small section of code notes: 'When you're concentrating on that little world you leave everything else out' (p. 145); and the project manager likens the project to a computer game: 'It's like being in Adventure. Adventure's a completely bogus world, but when you're there, you're there' (p. 95).

In *The Soul of the New Machine* the participants live in a world distant from 'ordinary life'. It is a world that asks little of its subjects other than that they be good computer engineers. It is a highly structured and seductive environment that may look bogus from the outside but that is forcefully real for its participants. One of its attractions seems to lie in its binary, black and white, character.

> The engineer's right environment is a highly structured one, in which only right and wrong answers exist. It's a binary world; the computer

> might be its paradigm. And many engineers seem to aspire to be binary people within it. No wonder. The prospect is alluring. It doesn't matter if you're ugly or graceless or even half crazy; if you produce right results in this world, your colleagues must accept you. (Kidder, 1982, p. 134)

The project manager was aware of the fragility of the parallel, perhaps even virtual, world he had created in the basement. He worried over the future of the team of young men he had developed once the computer they carried in their collective womb was from them 'untimely ripped' by the demands of the marketing department:

> The post-partum depression on this project is gonna be phenomenal. These guys [the team] don't realize how dependent they are on that thing [the machine] to create their identities. That's why we gotta get the new things in place. (Kidder, 1982, p. 205)

Making the machine, breathing life into it, nurtures a particular form of identity and way of life only so long as the corporation continues to build new computers and values the particular contributions made by each team member. Once the machine is made and shipped out the basement door a potentially terrifying vacuum is left. At this point having given birth to a baby they no longer control — the marketing people even change the computer's name as soon as it's finished — the team is potentially at a loss. Thus, it becomes vital to 'get the new things in place' and repeat the exhausting cycle again. At this point re-entering the earth's atmosphere and the world of the ordinary and mundane can be a terrifying prospect. So rather than rest and relax there is a compulsion to do it all over again, and again, and again.

This world is also vulnerable in that while being self-enclosed it is not self-sufficient. Its members still have to eat and sleep and interact with a more confusing and less predictable social world. And whereas the soul of the new machine judges on the technical merit of its engineers, the social world outside demands a more complex and contradictory presentation of self. In this external world binary reasoning is not enough; identity and masculinity have to be concocted and buttressed out of a changing and dynamic social relations. In particular, in the external world masculinity is perforce pushed back out onto the dance-floor in the confusing interchange with, and boundary work between itself and femininity.

Business Application Software: Masculinity and the Making of Software

In this part of the chapter I want to examine the way in which the work of software development is organized. This is because technology as an artefact and culture is not the only gendered aspect; the production and process of making technology is also gendered and if unchallenged strengthens links between prevailing conceptions of masculinity and making technology. Here, in particular, I want to question the apparent good sense of the prevalent method of developing business application software by dedicated project teams. I shall suggest that this particular form of organizing software work, and the vocabularies of motive employed in so doing, are closely related to prevailing masculine conceptions of work and technology.

Most application software development is organized in discrete projects (Friedman, 1989). There are considerable areas of software development that are not project-based, such as software maintenance, but these tend to be seen as low status and unchallenging areas of IS work. High status work tends to be associated with high profile new software development. It is prized for the inherent challenge it offers and the promotion prospects that follow the successful completion of major projects. Despite much debate about software quality assurance the success of new business applications software is still judged foremost by the ability of IS managers to meet system development time and cost constraints.

Project-based work has a particular culture and tempo that sets it apart from much routinized work. A male Project Manager defined the 'project mentality' thus,

> It is a different mentality. The mentality in admin is very much nine to five. Here, I mean my God, I come in at eight in the morning, I leave at seven in the evening, and there're still people here. It's a different mentality. If you ask people for a little bit of extra effort you get it.[2]

I asked this manager if he could have developed computer systems, with (women) clerical users in their offices.

> No, because at the end of the day we're still a project, at the end of the day we've still got to have a project mentality, you've got to run it [laughs] like a project; there must be very clear milestones, deliverables, objectives . . . which I think we would have lost sight of if we'd plopped the whole lot into the admin area.

So while project work is 'a lot of fun, hard work and a challenge' it also requires very clear parameters which make it visible and controllable. Put it in the administrative division and this visibility would have been diluted, dissipated, lost. Better that the technologists control it even if the rest of the company think they're 'weird fish'. Better to be 'weird fish' in your own pond than invisible in the admin area.

What is the 'project mentality'? It's the ability to give a little bit of extra effort, to work odd and often long hours and the possession of demonstrable competence in the discourse and techniques of 'milestones, deliverables, and objectives'. Increasingly, it's about having the right 'methodology' and being a 'software engineer'.

In business software development, project work means long hours. This is commonplace for Project Managers and Team Leaders, but programming and analyst staff are also expected to work long hours as projects near their release dates. According to Holti (1989) IS staff,

> strike a 'bargain' with organisational employment. They internalise an acceptance of the nature of project targets allocated to them in exchange for autonomy, lack of close surveillance and self-expression in the performance of their work. (p. 470)

Clearly working these hours can have benefits for individuals. If overtime is paid, already well paid staff can make considerable amounts of money. And where overtime is unpaid there is an expectation that there will be a payback in terms of promotion or preferential treatment. But for anyone with even minimal childcare responsibilities there are major problems.

Informal arrangements can be made for staff with young children up to a point. For example, one woman junior manager commented that she tried not to pressure people with families into working late or weekends 'unless it's absolutely necessary'. She said, 'We don't actually ask them [people with small babies] to come in if we can help it. [Pause] Well, I don't anyway. I don't think it's fair.' Often within the area of programming and analysis particular individuals will have skills and knowledge that are not easily transferable, however, a situation exacerbated by the tendency to allocate particular parts of programming or analysis to individuals. In this case either these staff do the work or the work may not get done. Staff with childcare responsibilities aren't 'actually asked', at least if their managers 'can help it', to work long hours, but as a woman senior software developer and single parent who read an earlier draft of this chapter commented,

> The fact that some people can live up to the expectation of long hours is nevertheless threatening to those who can't — a project I worked on got into difficulties and everyone was offered overtime incentives — although my manager accepted that I would not be taking advantage of them and quoted all the above stuff [regarding the company's respect for the individual] to reassure me that it was okay, I still felt almost guilty going home when the others stayed, particularly because almost everyone else was in a position to do so, which was stressful.

Even if work can be shared and some staff can work 'normal' hours this involves all sorts of hidden costs: the sense that you are not pulling your weight, that others are suffering for you, and that you are missing out on perhaps the most exciting and visible parts of the system development. For those with aspirations to get into the managerial grades in IS the inability or unwillingness to work long hours may be read as an insufficiency of organizational commitment.

If bouts of superhuman effort work against those who have other commitments or simply do not want to allow work to dominate their lives to this extent what do they do for those people who buy into or feel compelled to work in this way? In particular, what can we glean from these practices about the relationship between the development of a particular work-based masculinity and making software?

It is my impression that male IS staff and managers rather revel in the long hours they work. There is a tendency to glorify or accept as a technological inevitability the time they spend during the evening and at weekends at work. This separates out the IS man from the 'normal' business types; he might be a weird fish, and in the 1960s and 1970s he might have had long hair and strange clothes, but he could be relied upon to work long and unusual hours (see for example Pettigrew, 1973). The IS man, in his more unguarded moments, is apt to adopt a 'have a go' attitude, a 'we will deliver' mentality and a we'll-make-you-your-system-even-if-we-don't-have-the-resources approach that is simultaneously self-defeating and self-aggrandizing. This has a lot to do with the particular pressures IS men feel themselves to be under (see Knights and Murray, 1992).

There is a kind of *Boys Own* heroism about working these long hours. IS staff talk about preparing for the 'final push' and the 'muck and bullets' character of intense stages of project work. There is an ambiguous feel to the excitement that accompanies these phases of project work. An engineer in Kidder's (1982) study commented: 'It was a lot of fun, a lot of pressure' (p. 54); the project manager said, 'I'm flat out by definition. I'm a mess. It's terrible [pause] It's a lot of fun' (p. 109).

Project work appears to take on a life of its own; it is bigger than any of the individuals making it happen. You can either embrace it or take the difficult path of the conscientious objector, but in order to instil the project with glory and with a 'this thing is greater than us but we have to do it' dimension, frequent recourse is made to war imagery. The woman software developer who commented on an earlier draft of this chapter recounted an incident which graphically illustrates this point.

> A project I worked on got into difficulties and the lab manager set up a special project room with a label on the door of 'War Office'. Quite a few of us found this slightly ridiculous but nonetheless I know others enjoyed this sort of thing and the phrase 'blood all over the walls of the War Office' was a frequent one whenever a project manager got grilled over not meeting his dates. As a woman once active in the peace movement I found his attitude particularly alienating and there is no way I would have been able to 'buy into' it. The fact that a fair number of my male colleagues also found the whole thing puerile was an important factor in minimizing the tension at work.

From one perspective the comparison of software development with warfare is preposterous. Sitting in front of a workstation in well-appointed office accommodation in the the Home Counties in the 1990s is a world away from the 'muck and bullets' of trench warfare. From another perspective, however, these evocations of another theatre of masculine practice tell us something about the psychic reality of at least some men's experience of work. In this version, work is seen as war, as a matter of life and death struggle, of collective and individual heroism and sacrifice for an obscure and greater good.

War imagery can mobilize deep psychic energy in men. It also helps to make sense of the competitive social relations of capitalist work processes. This refers not only to the classical Marxist understanding of 'class warfare' but also to the warfare of inter-managerial and inter-specialist competition (see for example Jackall's (1988) account of US corporate alliance building and back-stabbing). Making sense of work through warfare and mobilizing energy through the evocation of war does something to the way work and masculinity are organized. Work becomes a dangerous and heroic struggle and the imagery and practice of a dominant masculinity mediates, shapes and personalizes market forces. And, as in warfare, so in work there are many casualties.

In the practice of software development particular pressures fall on Project Managers, Project Leaders and Team Leaders. Anecdotal evidence of the results of this stress on the health of junior IS managers suggests that the pressures are not to be taken lightly. Yet it is often only by succeeding and being seen to succeed as a Project Manager or Leader that the aspiring IS staffer can gain a foothold within the ranks of IS management. This clearly creates a dilemma for those who aspire to move into management. It also leads to the worst excesses of self-destructive macho behaviour.

One of the IS middle managers I interviewed talked surprisingly openly about his experience as a Project Leader on a highly visible project. His main recollection of the experience was being 'squeezed'. He said,

> Pressure definitely settles on the Project Leaders. The reason for that is you're not making the decisions . . . As a Project Leader I was receiving decisions and then you're squeezed; you've got close responsibility for the team under you who are also under pressure so you're bang in the middle where you're squeezed.

This manager has seen some of his peers move out of IS or into quiet backwaters of the department as a direct result of having had enough of this pressure. But despite believing that Project Leaders are made scapegoats and squeezed he wanted to have a go, to accumulate the right stuff in his climb up the ziggurat: 'You're trying to prove yourself, to get up to the next level.' Consequently this manager became a Project Leader and got his project in more or less on time. As the project neared completion, as it was born, however, his body gave up on him:

> I was probably keeping myself going during [the project] and then when I'd finished my body said, 'Forget it'. Management were very sympathetic. I dragged myself back into work for a week to do the budgets. I couldn't delegate it. The last two days I was told I was slurring my words.

When the tape recorder was off this manager recalled how he had been 'doubled up in pain' in the office during this period. Finally he went to see a doctor and was rushed to hospital. Nevertheless, he was back at work in a week to show, as he said, 'that I was OK'. After all, he said without a trace of irony, 'I'd never been ill'.

Having clawed his way back into the office to show he was OK this manager was then off work for over two months. In this period he had time to reflect and a number of consultants suggested to him that his worklife and his illness were probably connected. This led him to conclude that his illness was 90 per cent due to work stress. According to this manager his problems were created by the 'macho attitude in DP' based on the motto 'We will deliver'. Here though in order to fight his way up into management he ended up playing the very game he criticized and it was only when his body said 'forget it' that the began to reflect on where he was going. Still, when we spoke he seemed determined to continue to play this game in order to progress beyond the destabilizing position of Project Leader. Indeed, he felt the danger for someone like himself who had moved up the hierarchy pretty quickly was 'that if I stop at a certain level people will say I've reached my level of achievement'.

Putting the Pieces Together

Drawing on other people's and my own work this chapter has so far argued that there is a strong and 'naturalized' cultural connection between masculinity and science and technology. This is not surprising given masculinity's attempt to define itself by its monopoly control of reason, logic and objectivity. This symbiotic relationship of mutual interdependence has not been easily achieved. Rather, men have struggled to keep women and the feminine out of their masculine domains and when unsuccessful have attempted to ensure that women interlopers make a choice between their gender identity and membership of the science and technology 'fraternity'.

This relationship between masculinity and science and technology can be looked at from a number of different perspectives. Often these perspectives are seen as mutually exclusive and considerable hostility is generated between proponents of them. My view at the present time is that it is fruitful to adopt a multi-dimensional approach to the masculinity/science and technology relationship. As a tentative conclusion to this chapter I want to suggest a way of categorizing these different approaches to the masculinity/science nexus. This deliberately separates out these approaches in order to clarify their differences and their complementarities.

Following on from the different materials I have drawn on in this chapter it seems to me that there are at least three broad perspectives from which to examine relations between masculinity and science and technology. These are the cultural, the structural, and the psychoanalytical. These labels are very approximate and to be taken with a pinch of salt.

From a cultural perspective masculinity's enduring and changing relationships with science and technology are seen to be largely a cultural problem. It attempts to show us how science and masculinity have become intertwined and interdependent. It also challenges the naturalized cultural norms regarding this interdependence and identification. As such, the cultural perspective advocates a cultural battle, a kind of Gramscian struggle to undermine and change the hegemonic culture of technology, and its strong associations with masculinity, in contemporary society (as Henwood argues in this volume). This might take the form of the historical analysis of the emergence and changing character of these norms, as part of a larger project of exploring feminist approaches to science (Bleier, 1986) or focus on contemporary technologies and technological discourses as cultures that define and develop particular power/knowledge relations.

A structural perspective shifts the focus of analysis to look at the way scientific and technological occupations are gendered, where this includes the study of areas such as pre-job entry career choices, entry to trades and professions, and the structuring of work organization, culture and practice in those occupations. This approach explains how particular places are created within occupations and how these are filled and controlled. It advocates proactive campaigns for the entry of women into science and technology professions by the reorganization of training schemes and the removal of interior barriers to the promotion of women.

A common feature of both the structural and cultural perspectives is their concern with power relations. Thus the analysis of class and male power as capitalism and patriarchy are respectively given a central position in the theorization of gender relations.

A third perspective from which to look at the relationship of masculinity and technology is that which I have called the 'subjective'. By 'subjective' here I want to suggest those approaches that examine the subjective experiences that both cement and challenge links between masculinity and science and technology. These include broad psychoanalytic approaches, like that suggesting scientific endeavour as womb envy compensation, the kind of work undertaken around the social becoming of engineers and their individual biographies, and those studies that attempt to get beneath the surface of men's control of science and technology to look at the ways in which masculinity is attached to and identifies with technology.

This approach is concerned to examine the ambiguities of masculinity's partnership with technology at a symbolic and subjective level. This relationship is not straightforward, unitary or fixed in time. Nor do I believe that it is possible to portray it in terms of an omniscient male power protecting and reproducing its known interests.

From this perspective the mutual embrace of masculinity and technology is as much about the vulnerability of masculinity and men as it is about the conscious exercise of power and control. Through its apparent fixity and predictability, its reason and its objectivity, technology echoes and confirms a predominant masculinity's search for a controllable world of secure, certain and maintainable boundaries. Present technological cultures and practices also glorify and give credence to a masculinity alienated from emotion, the body, nurturing and domestic concerns. Technology legitimizes and makes 'natural' binary ways of thinking and an obsession with planning and certainty.

As such, technology as culture, as practice and artefact symbolizes and secures a separate reality of masculine perception, thought and action. The strong interlocked components of the technological world hold and stabilize masculinity's internal tensions and troubles. Technology becomes both a crucible and core domain of masculinity.

It must be said that the subjective perspective does little to address the issue of power in the relationship of masculinity and technology. However, its strength lies in its desire to understand the psychic and symbolic significance of technology for masculinity. I believe this is one of the crucial aspects of this relationship which needs further exploration, because masculinity's attachment to technology is not an adjunct or an optional extra. For masculinity, and perhaps particularly for technologists, technology is a core domain and discourse. Through this and other discourses masculinity affirms and makes sense of itself.

The practical significance of this perspective is perhaps most apparent when we look at male resistance to the 'dilution' by women and/or the feminine of science and technology as culture and practice. Resistance springs not just from a protection of power and privilege. I would suggest it also comes from a deeper motive to protect a masculine reality that has secured itself in the symbolic and processual significance of science and technology.

In other words, to 'take the toys from the boys' threatens those boys with the removal of one of the symbols that makes them feel like boys and, significantly, not girls. Without those 'toys' (the whole array of technological artefact and culture) the boys would no longer be boys as they and we know them.

The privileging of the subjective perspective here is not intended to marginalize the structural or cultural approaches. I see them as complementary in a difficult and non-continuous way (see Craib, 1989). Nor is highlighting the vulnerability of masculinity and its now necessary attachment to technology a plea to leave men or masculinity alone. On the contrary, my purpose here is to try and understand how men can lessen the interdependence of masculinity and science and technology in such a way that the annihilation threatened to a masculinity devoid of its technological twin is countered or supplanted by a rosier and more promising future.

Notes

1 This chapter has been through a number of drafts. Comments from readers and colleagues have been of great help. I would particularly like to thank the editors of the present collection, Ardha Best, Doug Bell, Janet Low and the woman software worker who read and commented on the chapter. (The latter has asked to remain anonymous.) Needless to say, the stumbling and pratfalls are my own.
2 The material presented here draws on research in the financial services industry. In this my colleague, David Knights, and I examined the political dimensions of systems development (see Knights and Murray, forthcoming). As the research developed I also became increasingly interested in the way in which systems developers see their work and organizational worlds (see Murray, 1991).

References

BLEIER, R. (Ed.) (1986) *Feminist Approaches to Science*, Oxford, Pergamon.
COCKBURN, C. (1983) *Brothers: Male Dominance and Technological Change*, London, Pluto.
COCKBURN, C. (1985) *The Machinery of Dominance: Women, Men and Technical Know-How*, London, Pluto.
CONNELL, R.W. (1987) *Gender and Power*, Cambridge, Polity.
CORNEAU, G. (1991) *Absent Fathers Lost Sons: The Search for Masculine Identity*, Boston, Shambhala.
CRAIB, I. (1989) *Psychoanalysis and Social Theory: The Limits of Sociology*, Hemel Hempstead, Harvester Wheatsheaf.
EASLEA, B. (1983) *Fathering the Unthinkable: Masculinity, Scientists and the Nuclear Arms Race*, London, Pluto.
FRIEDMAN, A. (1989) *Computer Systems Development: History Organization and Implementation*, Chichester, Wiley.
HACKER, S. (1990) *Doing It the Hard Way: Sally L. Hacker*, (edited by D.E. SMITH and S.M. TURNER), London, Unwin Hyman.
HARDING, S. (1986) *The Science Question in Feminism*, Milton Keynes, Open University Press.
HOLTI, R.W. (1989) 'The Nature of the Control of Work in Computer Software Production', unpublished PhD Thesis, Imperial College of Science and Technology, London.
JACKALL, R. (1988) *Moral Mazes: The World of Corporate Managers*, New York, Oxford University Press.
JORDANOVA, L.J. (1987) 'Gender, Science and Creativity', in MCNEIL, M. (Ed.) (1987) Gender and Expertise, pp. 152–7, London, Free Association Books.
KIDDER, T. (1982) *The Soul of the New Machine*, Harmondsworth, Penguin.
KNIGHTS, D. and MURRAY, F. (1992) 'Politics and Pain in Managing Information Technology: a Case-Study from Insurance', *Organization Studies*, 1992, **13**, 2, pp. 211–28.
KNIGHTS, D. and MURRAY, F. (Forthcoming) *Markets Managers and Technology*, Chichester, Wiley.

MURRAY, F. (1991) 'Technical Rationality and the IT Specialist: Power, Discourse and Identity', *Critical Perspectives on Accounting*, **2**, 2, pp. 59–81.

PETTIGREW, A. (1973) *The Politics of Organisational Decision-Making*, London, Tavistock.

ROSSITER, M. (1982) *Women Scientists in America: Struggles and Strategies to 1940*, Baltimore, John Hopkins University Press.

SEIDLER, V.J. (1989) *Rediscovering Masculinity: Reason Language and Sexuality*, London, Routledge.

WOLFE, T. (1980) *The Right Stuff*, London, Picador.

Chapter 5

The Expert Systems Debate: A Gender Perspective

Alison Adam and Margaret Bruce

In this chapter we explore the many issues surrounding expert systems and gender. The scope of our concerns is very broad ranging, perhaps more so than with any other branch of computing. This is because we must consider not only the pragmatic matter of women designing and working with expert systems but also more philosophical questions relating to the type of knowledge or expertise represented in such systems, in other words, knowledge representation. In addition to this we also consider what women themselves feel about the likely impact of expert systems on their lives. Before discussing these issues we locate expert systems work in relation to the history of artificial intelligence.

Introduction: A Short History of AI

Artificial Intelligence (AI) is the branch of computing which is concerned with simulating intelligent human behaviour by machine. Although there is no consensus as to a precise definition among practitioners of the discipline, most would accept that it encapsulates a two-fold view of intelligence. On the one hand AI attempts to simulate intelligence but on the other hand it also offers models of cognition in computational terms. AI has potentially wide ranging applications across all areas of human activity from managerial decision-making to geological and medical expertise.

AI is as old as modern computing itself. The earliest efforts in machine intelligence were made in the 1950s when game playing, especially chess playing programs were produced and there was much interest in developing systems which would model general problem solving abilities. In the latter area, one of the best known programs is GPS (General Problem Solver) which was only effective on severely bounded domains such as small logic problems and artificial toy block worlds (Newell and Simon, 1963). The mood of the time was one of optimism, even bordering on arrogance.

> They were interested in intelligence, and they needed somewhere to start. So they looked around at who the smartest people were, and they were themselves, of course. They were all essentially mathematicians by

> training, and mathematicians do two things — they prove theorems and they play chess. And they said, hey, if it proves a theorem or plays chess, it must be smart. And they found out that, for a number of reasons, this really seems to miss the point .... The point about people is how they do the easy things, not how they do the hard things. (Solomonides and Levidow, 1985, pp. 18–19)

Later on, in the 1960s, it became clear that the relative failure of the generalized problem solving systems of that period was due to an excessive emphasis on a view of intelligence in terms of problem solving at the expense of knowledge of a particular domain whether medicine, engineering or whatever. Needless to say AI suffered a decline in fortune in the early 1970s, and nowhere more so than in the UK. It had promised much but delivered little and was to languish in the doldrums until the end of the decade when the Japanese announced their fifth generation computing project. The British followed in their wake and implemented a five-year research programme unprecedented in both size and scope. It was partly state funded through the Department of Trade and Industry and partly funded by British industry. The remit of this programme, the Alvey project, named after the chairman of British Telecom who instigated the initial report, was to investigate a number of enabling technologies ranging from Software Engineering to Intelligent Knowledge-Based Systems (IKBS) (Alvey, 1982). After the end of Alvey the mood, both in industry and the academic world remains ambiguous as to the success of the project as a whole. Yet success or failure is not the point at issue here. What Alvey did do was to create a number of jobs, albeit mostly temporary research posts in AI related fields and also to generate a terrific upsurge of interest in AI in general. AI is now taught in most university computing courses; this was not true ten years ago. In addition to software houses specializing in AI, many large organizations, whether in finance or chemicals, which support a computing research function now undertake research and development into areas of AI.

It is the applied end of AI which has been of most interest to industry in the shape of IKBS, a term which is often used synonymously with ‘expert systems’. If there is a distinction to be made between the two terms it is that an expert system explicitly models the expertise of an expert or experts in some applied domain, for example, the knowledge of a medical practitioner, scientist or accountant, whereas an IKBS can be designed which does not explicitly model the reasoning processes of an expert. In other words expert systems can be seen as a subset of IKBS. Additionally we should have high expectations of the performance of expert systems. They should perform at or near expert level in terms of accuracy and speed of diagnosis or decision. We should expect an expert system to have a good user interface, in other words the part the user ‘sees’, which consists of not only the layout of the screen but also how the system accesses and handles knowledge, must be natural and usable by an individual who may not be an expert. This is especially true since the final product is to be used in a real setting as opposed to an experimental or research environment. Finally we should expect an expert system to offer an explanation of the decision or advice offered since that is what we would expect of a real expert. A more experimental IKBS, perhaps designed to test some hypothesis about human decision making, and perhaps only used in a research environment, need not be bound by these more pragmatic concerns. The final distinction is cultural. One is far more likely to hear the term 'expert system' in an industrial setting, while IKBS is more common currency in academic and research environments.

If expert systems can be seen as at the modern end of AI, they are also at the practical or applied end of AI. Gone are the heady days of the late 1950s and early 1960s where it seemed that the turn of the century would witness the dawn of the truly intelligent machine. The claims and aspirations of present day practitioners of the discipline are more modest and realistic.

What Are Expert Systems?

The emphasis of expert systems research lies in the notion that intelligent behaviour is rooted in domain knowledge rather than domain independent problem-solving ability. This means that expert systems are designed with an explicit symbolic representation of knowledge. Such a knowledge base is at the heart of an expert system and is essentially more important than a clever problem-solving algorithm. This mirrors the idea that an expert is an expert because of what s/he knows rather than because of problem-solving abilities although it would be wrong to suggest that the two are completely divorced. The difference is rather like that of the student at the beginning and end of a computer science degree. At the beginning the student has a set of problem-solving skills learned in childhood, at school and so on but perhaps little by way of domain knowledge. By the end of the course these elements of intelligence are still there and new problem-solving skills may have been learned but these have now been augmented by a large amount of domain knowledge acquired over the three years during which the individual has been studying. However the real point about expert systems is that they are a way of capturing, to some extent, expert knowledge, and in our society that usually means male knowledge.

There are a number of different knowledge representation formalisms ranging from those based on formal logic to those based on a model of how knowledge is stored in human cognition. Although exponents of logic based formalisms point to the precision and generality afforded by logic, it is clear that most logics have no satisfactory way of dealing with structural relationships such as hierarchies and inheritance properties within hierarchies. For instance if you were producing a taxonomy of animal types you would want the 'bird' category to inherit most of the properties of 'animal' such as 'has skin' and 'can breathe' but in addition the bird category has the properties of 'has feathers' and 'has wings'. Now canaries are a subset of birds which would inherit all the general bird and animal properties but in addition we would like them to have the properties of 'is yellow' and 'can sing'.

In addition to a knowledge base, an expert system must have a means of reasoning with that knowledge. The part of an expert system which deals with this is usually termed an 'inference engine' and as we shall describe in the next section it is no accident that engineering terminology runs through the language of expert systems. This is really the problem-solving side of the system and once again there are a number of techniques commonly used.

There are a number of differences between current-day expert systems and older-style general purpose AI systems but the most important distinguishing element must be the fact that both the type of knowledge representation and design of the inference engine depend heavily on the domain under consideration. However there is no generally accepted classification of domains of knowledge within the expert systems world and it seems as if it is not always easy to translate ideas used in one domain to another domain. For instance different scientific domains are very different from one another. Legal knowledge and medical knowledge must be represented differently and common-sense knowledge, shortcuts and strategies must be represented.

The other important parts of expert systems include the user interface and the explanation sub-system. In computing in general, much research has been undertaken on the design of user interfaces and many ideas can be imported wholesale into the expert systems world. Indeed it may be preferable not to distinguish the user interface as something separate from the underlying system which can somehow be bolted on afterwards. The main additional consideration for expert systems probably lies in the dialogue which may be more complex than for many traditional systems. For instance at some points the system may query the user for information, while at other times the user may question the system (Hammond and Sergot, 1983).

Explanation is still very much a research topic in the expert systems world. At the simplest level an explanation can be generated which is simply a trace of everything that the system has done to get to where it is at the moment. Such a trace can be of use to the system designer in ironing out bugs in the system's reasoning and is also probably better than nothing for the user as long as the system translates its trace from computerese into English but it does not really constitute an explanation in semantic terms (Jackson, 1990, p. 334). The problem here is that if we are required to trust expert systems then we must have more sophisticated explanation facilities.

Philosophical Explanations — Whose Knowledge?

When you come across expert or knowledge based systems for the first time you may be struck by the apparent strangeness of the terminology. Some organizations term their designers of expert systems 'knowledge engineers' as if knowledge can somehow be engineered in the same way as a network of pipes or a car engine. In other words knowledge seems to be treated as something material and substantial rather than something mysterious and intangible — for engineers rather than philosophers. We argue that the widespread introduction of artificial intelligence in general, and IKBS in particular could imply the acceptance of a materialist epistemology. We mean 'materialism' in the old Victorian sense which saw the world and all its contents in terms of forms of matter. The research problems of acquiring and representing knowledge, interesting and important though they may be, are not really the points at issue here. It is instead the ideology surrounding the technology which is important — the ideology which allows you to smile the first time you hear the term 'knowledge engineer' yet means you accept it unquestioningly within a short period of time.

Whatever the technical problems, expert systems technology implies a belief that it is possible to extract the knowledge individuals have and make a true representation of it in a computer system so that the system can reason in the same way as or perhaps even better than the individual. The knowledge has to be made explicit, formalized and so, in a sense, material. Making knowledge material in this way has a number of potentially serious implications.

First of all there is the question of whose knowledge is to be formalized in this way. Which individuals or groups will this apply to? It is tempting to suggest that it will be the knowledge of groups in power and that intelligent knowledge based systems will in fact be white middle-class male knowledge based systems and will exclude the knowledge of ethnic groups, the disabled and women. Our discussion is obviously based around the latter category but may apply analogously to the other groups identified.

One of the important strands which has appeared in women's studies literature in recent years relates to the process by which history has obscured and forgotten what is peculiarly woman's knowledge. Eherenreich and English (1979) have described the rise of the male medical expert at the expense of female lay healers or 'wise women'. Never was the old adage 'knowledge is power' more true. But the power of the knowledge is not a neutral thing, rather it is the value which society places on that knowledge and this is something which changes over time. Nevertheless historical studies demonstrate that when a particular skill becomes associated with male rather than female activity it acquires additional status in the eyes of society. In the medical example the process was a slow one but there are other examples occurring over shorter timescales. For instance most of the earliest computer programmers were women and the job was thought of as a clerical or calculating role (Giordano, 1988). Quite quickly men began to join the ranks and the job began to be seen as a skilled professional job. Although the tasks that computer programmers actually do have altered a great deal over forty or so years the level of skill required has always been substantial and has neither increased nor decreased even if the form of the skill has changed considerably. However the job tended to be perceived as a skilled one when it was taken over as a male preserve.

Although women may join male professions such as medicine or programming, professions where women predominate are always accorded lower status. Much of the equal opportunities legislation introduced in the UK over the last fifteen to twenty years has attempted to redress this balance. Cases may be brought to court under the Equal Pay Act to determine whether an individual's job is of equal value to another's against which it is being compared. As it is impossible to state exactly what 'equal value' actually means it is unlikely that the status of women's jobs will change through this route, and on a more practical level that women will be paid substantially more for doing the same job. As legislation cannot change social attitudes the work which is done by women and hence their skills and knowledge is unlikely to rise in value in the eyes of society.

It seems likely that expert systems will reinforce rather than undermine the low value which society places on women's knowledge. One of the reasons relates to the emphasis on the expert. Experts in our society are usually professional men and current expert systems research does little to challenge our reliance on expert knowledge. Indeed the traditional good advice for choosing an expert systems project, suggests looking for a domain with a clearly identifiable single expert rather than an area where there are no experts or several competing experts (Welbank, 1983). This emphasizes the individualistic nature of expert knowledge and focuses attention away from the idea of knowledge as a shared resource among a group.

The additional authority vested in expert knowledge by way of its becoming formalized and reified may be undesirable for several reasons. First of all it serves to mystify an area of knowledge rather than make it accessible. Second, it may exercise an undesirable normative effect in relation to that expert knowledge. It is possible to imagine a scenario where it is easy to be intolerant of mavericks and pluralistic discourse is discouraged. Expert systems could well be used to bolster the economic and social views of political systems, make 'facts' out of 'views' and thereby discourage challenges to authority.

The second issue which we wish to raise in relation to the materialist epistemology of expert systems relates to the question of language and the metaphors used in artificial intelligence work. Modern computing has designated itself a branch of engineering. We have software engineering, information engineering and, as we have already suggested in relation to knowledge based systems, knowledge engineering. The philosophy behind this is that principles and practices which belong to more traditional branches of engineering should be imported into the production of software and information systems for managers and professionals. This involves the development of techniques to establish good standards of design, reliability, quality and testing. In the case of software systems, such engineering standards should make for more reliable and therefore more maintainable and ultimately cheaper software. Certainly these are overt reasons for adopting engineering principles but there are some less obvious reasons why computing should find a home in engineering. Although some of the older members of the profession were electrical engineers rather than mathematicians, for most computing professionals their work does not fall neatly into a science, a technology or a business activity. The banner of engineering offers some measure of prestige and a professional home to a new hybrid discipline although not unambivalently, as in the UK, at least, the overall status of engineers is lower than their Continental counterparts. Engineering may then be seen as the professional refuge of the marginal men of computing.

Artificial Intelligence research has its roots both in cognitive psychology and in traditional computing so it is not surprising to find that the engineering metaphor spills over into this domain. Engineering involves the construction of material artefacts. If there is some difficulty imagining the engineering of software, it is even harder to visualize the engineering of knowledge. In order to do so knowledge must be treated as a material entity. The use of the engineering metaphor becomes clear when we examine the language of knowledge based systems. Many IKBS are constructed in the form of a knowledge base composed of rules. The mechanism which reasons with these rules is termed an inference engine. As the inference engine proceeds in its inference cycle, rules are 'fired'. The system is analogous to a great engine with rules firing away like pistons and the knowledge engineer stoking the boiler with knowledge. For instance MYCIN is a well known example of a medical expert system where the object of the system is to suggest the nature of an infectious organism and an appropriate treatment (Jackson, 1990). Rules are of the form: 'If the strain of the organism is x and the morphology of the organism is y then there is suggestive evidence that the organism is z'. Using the clinical evidence available for a particular case, rules are fired in a backwards sequence, working backwards from the potential goal of a candidate organism and using observed data or data supplied by the physician to establish the identity of the organism with certainty.

On its own the engineering metaphor is unlikely to appeal to women who have traditionally been alienated and excluded from engineering disciplines. More importantly mechanistic views of human intelligence tend to reinforce this exclusion as it is only possible to promulgate such mechanistic views of human intelligent behaviour by ignoring or denying alternative views which stress the non-rational nature of intelligent human behaviour and the social and cultural nature of knowledge. The difficult non-mechanistic side of human intelligence includes not only intuition and creativity but skills which may belong to a particular group. In other words traditional female caring skills, part of the 'feminist epistemology' which Hilary Rose has proposed can be forgotten or ignored (Rose, 1987). This means that expert systems may be used both to reinforce traditional notions of expertise and experts and additionally may emphasize mechanistic views of human expertise. Yet expert systems do not have to be used in this way if women are involved in their design and creation.

Who Designs the Expert Systems?

It is common knowledge that the percentage of girls and women entering higher education courses in computer science and computing continues to decline. The overall percentage of women on such courses had declined from about 25 per cent in the late 1970s and early 1980s to 10 per cent or less (Lovegrove and Hall, 1987). In the face of a persistent technological skills crisis and despite the launch of national initiatives the outlook is not optimistic. In discussing the shortfall of women in computing there is a danger of treating information technology as a unified discipline without exploring the real differences which exist between types of computing jobs. Although something is known about women working in the more traditional areas of systems analysis and programming, job opportunities have only recently been created in artificial intelligence and knowledge based systems. On the one hand this is due to projects such as the Alvey initiative which created awareness of the potential of AI and established a pool of skilled researchers across the range of 'fifth generation' computing. In AI alone the number of academics increased from about two dozen to between 1,650 and 2,000, although most of these posts were temporary positions. On the other hand, the industrial world has woken up to the possibilities inherent in expert systems and it seems likely that the job market for AI will remain steady despite the recession of the early 1990s.

In a recent study of the expert systems industry which included interviews with almost all of the software houses in the UK producing IKBS software we found only one women working in knowledge engineering (Bruce, 1988). It is worth exploring why this is the case. Research in the design industry (Bruce and Lewis, 1990) has identified the barriers or hurdles which affect women's career paths. We believe that this model can be broadly applied to several different professions and is particularly suitable for describing women's careers in the AI industry.

This model identifies three barriers or hurdles acting at critical points where women make decisions about their careers. These are, 'getting qualified', 'getting a job' and 'getting on in a job'. At the first hurdle educational courses are seen as 'technical' and therefore difficult for women. At the next stage the stereotyping of jobs can discriminate against women and finally the ability to get on at work is influenced by women's perception of themselves as essentially task rather than career oriented implying that they fail to work in ways which would help to augment and develop their careers.

The first stage of getting appropriate qualifications in AI raises some interesting questions. Even before reaching the stage of further and higher education the stereotyping of computing as a male subject has begun at school. This means that few young women go on to study computing after school. But although there is a bias towards scientific and technological subjects, the computing industry recruits from all academic areas and also recruits individuals who have not participated in higher education. In fact many firms who could be described as computer users, rather than software houses and computer manufacturers, employ numerate graduates and train them as systems analysts and programmers.

To work in AI a first degree in computing or at least a conversion MSc is generally necessary. Apart from psychologists (because of the particular relationship between AI and cognitive psychology), it is difficult for graduates of other disciplines to move into AI and even more difficult for non-graduates. With the small percentage of women entering computing courses the pool of qualified women who could contemplate a career in AI is much smaller than the number who could potentially enter the computing industry at large.

The second phase of 'getting a job' is another hurdle to cross. AI jobs may be seen as more glamorous than more conventional jobs in computing and can be regarded as promotion for 'high-flyers'. It is unlikely that the first work experience for graduates will be in AI. This means that a well qualified woman who wanted to move into AI would have to gain appropriate work experience first, for example with a blue-chip company or a reputable software house. It would then be necessary to apply for a job and to overcome the lack of confidence which many women feel in attempting to move into what is seen as a highly technical area.

The third hurdle of 'getting on' in a job is related to the second. If AI groups within organizations consist of dynamic high-flyers, this may go some way to explain why women are not generally visible in the AI world. Women tend to be overly 'task-oriented' rather than 'career-oriented' thus gaining excellent technical skills but without paying due regard to other aspects of work which may be just as, if not more important, to getting on in a career. Additionally if it is true that a move to AI could come later in a women's career then there is the perennial problem that the arrival of children may come at a time when career pressures are at their greatest. A career break at this point may mean that it is impossible to keep up with a rapidly changing technology and difficult to re-enter a profession where the age profile is slanted towards individuals in their twenties and early thirties.

So far, an expert systems industry does not exist as something separate from the computing industry at large and there may never be a neat segmentation of the marketplace. Nevertheless we suggest that, for the reasons already outlined, women may find it even more difficult to get on in the world of AI than in more conventional areas of computing. The reasons we have described are not new and may be just as applicable to other professions but they do suggest that women as designers and creators of expert systems are likely to remain very much in the minority even if the balance shifts substantially in other areas of computing. This means that there is unlikely to be a challenge to the orthodox notions of expertise and the focus on the expert which is offered by the expert systems world from this quarter.

Women's Views of Expert Systems

As knowledge engineering is a specialized profession, relatively few individuals, whether men or women, are likely to be involved in the design and creation of expert systems. Yet many of us will be affected as users or what might be termed 'subjects' of expert systems. Traditionally in the world of industry, users are the individuals or groups of individuals from various work and management functions who have commissioned a computer system which is designed and programmed by the data processing function within an organization. Such users have a great deal of power in relation to the systems which result and much of the systems analysis literature is geared towards the need to establish user requirements, to design good software interfaces and so on (Hekmatpour and Ince, 1986).

However many of us are users of computer systems over which we can exert little or no control. Anyone who uses an automatic cash machine or a library catalogue system is also a computer user but in this case has no say in the system design and must put up with a machine which won't dispense a bank statement or must accept a verbose search through a series of references. At best, systems designed for the general public in this way can be efficient providers of information (and perhaps cash!); at worst they can be irritating and restrictive.

Hence many individuals who would not perhaps regard themselves as computer users have unwittingly accepted the use of computers in their lives. But even those of us who never directly use computers are the subject of some computer system somewhere. In fact almost every adult in the UK has a data record ranging from banking, payroll, Department of Social Security records to catalogue mailing lists and so on.

Many of these considerations apply to expert systems. When computer systems work as they are intended to we take their usefulness for granted. When they do not work as they are intended to, however, we may find ourselves in situations where we are powerless to influence the way they act and where it is difficult to obtain redress where, say, we have been refused credit. It is tempting to think that men and women are, at least potentially, equally affected as subjects of computer systems and that no gender questions are raised, but this glosses over the different reactions to authority which are experienced by men and women. As individuals, women feel powerless in the face of the authority of officialdom. The use of computers does nothing to alleviate this. Instead computerization may make working practices and much information set in concrete. Women who have little access to technological information are likely to feel much less able to challenge such practices. The powerlessness which women experience is likely to be exacerbated by expert systems precisely for the reasons we have explored above. If women themselves are the subjects of expert systems then opinion and belief can become knowledge and that knowledge can become fixed and immovable if it is contained within an expert system. In some contexts the results may not appear sinister, as in a system for providing tax advice or for geological prospecting, but there may be more serious implications involved in social security or medical expert systems.

Whether they like it or not women cannot ignore the effects of computerization and, if expert systems become widespread, cannot opt out of their use. However there is evidence to suggest that reactions to the possible use of expert systems is not uniformly negative but ranges from hostility to qualified optimism. For instance we had the opportunity to carry out an assessment of the likely impacts of expert systems on women's lives with two different groups of women (Bruce and Adam, 1989). The first group were women who were professionals in the computer industry or in higher and further education while the second group were women on a government training course. The second group were on one of the many schemes which proliferated in the 1980s under the heading of employment training for young people. They had left school with few qualifications and were unemployed when they started the scheme. Not surprisingly the views of the two groups contrasted sharply. The first group were worried about the deskilling of women's jobs but felt that in the long term expert systems would find applications in traditional areas of male employment and so men would be similarly affected. They suggested that expert systems could give the appearance of making more information available while subtly supporting the *status quo*. This group had lots of ideas as to how expert systems could be applied, many of which were very positive. They suggested that expert systems could even be used to counter sexism — for instance an expert system for career guidance would not be able to see whether its subject was male or female.

By contrast the second group's response was uniformly hostile. Their fear of the technology was very evident. They saw computers as malevolent boxes. Job opportunities would be lost and, for them, expert systems meant isolation and loss of work. Yet on the other side they could see places where expert systems, or similar technology, could be used positively. For instance in the home they were willing to imagine household robots and super-efficient houses. They could imagine efficient and safe cars designed by expert systems technology. Although not at first obvious to them, it is clear that their voices could be heard in spheres like these.

The group of professional women were more able to question the validity of the technology and its actual utility. They were ambivalent about its actual social impacts and could see that it could work either for or against women although it was likely to do the latter in the contemporary social context. The trainees were, however, extremely negative about technology, fearful and almost violently opposed to it. The first group of women were, on the whole, confident about using technology and clearly felt that they were in a position to make choices. They did not regard technology as 'out of control' but felt rather that it is necessary to debate and think through the social and political issues surrounding technological change. The group of trainees felt strongly that they would never be able to control or influence the design of the technology or its path of development. The group itself pointed out that access as a consumer to technology was related not only to gender but alto to socioeconomic class. This is the key issue in relation to women as designers, users and subjects of expert systems. Obviously the relatively few women who will be in a position to design and influence the development of expert systems will be drawn from middle-class professional backgrounds. However if expert systems become widely used the majority of women will, as the subjects of expert systems, feel as powerless as the group of trainees to understand let alone have an influence over the technology.

Conclusion

In this chapter we have suggested that the introduction of expert systems cannot be regarded as neutral with respect to gender. The reification of expert knowledge will tend to exclude alternative world-views, especially women's knowledge. The dearth of women working in the area is likely to continue despite attempts to increase the number of women entering computing as a career. For the majority of women there is little opportunity to influence the design of expert systems, despite the fact that they will be the subjects of expert systems. Because expert systems are not as yet widely used, many of the issues we have highlighted are still to be resolved in the future. Thinking of different futures with expert systems gives another dimension to technological change and highlights fundamental questions as to whether the technology is needed, and what options there are with respect to how it may be developed. Expert systems could be used to disseminate information and give impartial advice. They could be designed to be free from obvious prejudices and make more opportunities available for women. They could perhaps even relieve the domestic burden which most women have. The fact that there are alternatives means that the future of expert systems need not be entirely pessimistic in relation to gender issues.

References

ALVEY, J. (1982) *A Programme for Advanced Information Technology. The Report of the Alvey Committee*, London, HMSO.

BRUCE, M. (1988) 'New Product Development Strategies of Suppliers of Emerging Technologies — A Case Study of Expert Systems', *Journal of Marketing Management*, **3**, 3, pp. 313–28.

BRUCE, M. and ADAM, A. (1989) 'Expert Systems and Women's Lives: A Technology Assessment', *Futures*, **21**, pp. 480–97.

BRUCE, M. and LEWIS, J. (1990) 'Women Designers — Is There a Gender Gap?' *Design Studies*, **11**, 2.

EHERENREICH, B. and ENGLISH, D. (1979) *For Her Own Good: 150 Years of the Experts' Advice to Women*, London, Pluto.

GIORDANO, R. (1988) 'From the Frontier to the Border: Women in Data Processing 1940–1959', *Proceedings of the Joint Conference of the British Society for the History of Science and the History of Science Society*, Manchester, UK, Unpublished pp. 357–64.

HAMMOND, P. and SERGOT, M. (1983) 'A PROLOG shell for logic based expert systems', *Proceedings of Expert Systems 83*, Surbiton, M. Ashill, pp. 95–104.

HEKMATPOUR, S. and INCE, D. (1986) *Rapid Software Prototyping*, Milton Keynes, UK, Mathematics Faculty, Open University, Technical Report 86/4.

JACKSON, P. (1990) *Introduction to Expert Systems*, Wokingham, UK and Reading, MA, Addison-Wesley.

LOVEGROVE, G. and HALL, W. (1987) 'Where have all the girls gone?', *University Computing*, December, pp. 207–10.

NEWELL, A. and SIMON, H.A. (1963) 'GPS, a program that simulates human thought' in FEIGENBAUM, E.A. and FELDMAN, J. (Eds) *Computers and Thought*, New York, McGraw-Hill.

ROSE, H. (1987) 'Beyond masculinist realities: A feminist epistemology for the sciences', in BLEIER, R. (Ed.) *Feminist Approaches to Science*, New York, Pergamon, pp. 55–76.

SOLOMONIDES, T. and LEVIDOW, L. (Eds) (1985) *Computers as Culture*, London, Free Association Books.

WELBANK, M. (1983) *A Review of Knowledge Elicitation Techniques for Expert Systems*, Ipswich, UK, British Telecom Research Laboratories.

Section III

Gender, Clerical Work and Information Technology

Chapter 6

Information Technology and Occupational Restructuring in the Office[1]

Sonia Liff

Introduction

The introduction of new technology into an area of work as sexually segregated as the office raises important questions about the ways in which gender relations are reproduced and restructured through periods of change. Drawing on a study of women workers in the West Midlands region of the UK, this chapter explores the ways in which current technical change in the office has affected the content of women's work, and hence the experience of doing such jobs, without significantly disrupting the boundaries of gendered occupations. In seeking to explain this the chapter will explore two aspects of office systems design: first the scope of the technology introduced, and second the ways in which the systems are being implemented.

The consequences for women of the perpetuation of a gendered occupational structure are profound since women are largely confined to those levels offering limited pay, benefits and opportunities. In a detailed analysis of the 1984 Women and Employment Survey data, Dex (1987) found only 11 per cent of women engaged in professional or semi-professional careers. Twenty-eight per cent of women were predominantly employed in clerical occupations throughout their working lives. Dex comments:

> Of those who started out in clerical profiles, over 50 per cent were still in clerical jobs in their final working period, although there were many who were not in clerical jobs later on, most of whom had experienced downward occupational mobility. . . . Upward mobility was less common although it did occur mainly into the category called 'other intermediate non-manual' or into nursing and very occasionally into professional occupations. (p. 49)

Theorizing the occupational structure as gendered goes beyond observing patterns of segregation. It also involves analyzing forms of work organization and workplace relations as gendered. Studies of the office have taken such analyses further than other areas of women's work, initially by employing analogies based on husband/wife relations. Pringle's (1989a) more sophisticated analysis describes a variety of domestic roles, including those of mother and daughter, that can be assumed by the secretary and interestingly discusses the form these dynamics take when the manager is a woman.

Some writers argue that it is precisely these distinctive forms of work organization that make the office so resistant to change. For example, Bevan (1987) claims that opportunities for job redesign are blocked by 'the structural and social conventions surrounding the "feminisation" of office work and the control and power relationships which dominate it' (p. 186). This suggests that the status associated with having a secretary (and in line with this analysis one might argue the reinforcement of male gender identity), is such that managers are very reluctant to relinquish this form of work organization. Kanter (1977) does not see the barriers as just one-sided. She says that the secretarial function represents 'a repository of the personal inside the bureaucratic' (p. 101), and as such is valued by both managers and secretaries. Mannell (1987) claims that the problem arises because the organization treats a secretary and her manager as inextricably linked:

> In many organisations the secretary herself has little or no status at all and does not even appear on the organisation chart. To be a secretary is to be employed in an essentially supportive role, however senior is the principal. (p. 45)

Thus the job is not even seen as a thing in itself. Despite this, as Webster argues in Chapter 7, secretaries carry out a very wide range of tasks, often using considerable discretion.

The likely effect of technical change on these patterns has also been considered. In line with the arguments above, Pringle (1989b) suggests that the gendered relations of technology are such that male managers will be unwilling to hand over control of personal computers to their secretaries. In contrast, the early account by Barker and Downing (1980), while agreeing that the office was still governed primarily by patriarchal relations, argued that the significance of information technology lay in its ability to transform these to capitalist relations.

However it is more common for discussions of technical change in the office and elsewhere to treat technical and sexual divisions of labour as essentially separate dynamics, (Liff, 1987). This leads to discussions about the likely implications of technical change for the job content and work organization where these are seen as varying little with the sex of the worker. It also means that the scenarios presented are not generally informed by an analysis of gendered relations. There are three main approaches in the literature: first, the labour process perspective which suggests technical change will lead to reduced job opportunities and skill levels and increase the control exercised by managers over the workforce; second, more optimistic commentators who suggest that new technology will provide women with new skills enabling them to advance their careers by entering previously male-dominated occupations currently plagued by skill shortages; third, there is the possibility of a restructuring of existing divisions of labour which might transform the boundaries between men's and women's work. For example some writers suggest that there is potential for information technology to lead to less rigid boundaries between clerical, administrative and managerial work (Armour, 1986; NEDC, 1983).

Technical change in the office has been extremely widespread and provides a good site for exploring these dynamics. Daniel's (1987) analysis of the 1984 Industrial Relations Survey suggested that in the period 1981–1984, 35 per cent of office workplaces introduced computers or wordprocessors. Over a third of workers at such an establishment were in general affected by the most recent specific change, and it seems likely that these trends are continuing. However such studies say little about how women and gendered work relations will be affected by such changes.

This chapter attempts to combine analyses of technical change and gender relations in the office. Using material from a range of existing studies and findings from a study of women workers in Great Britain's West Midlands region, all three of the scenarios outlined above will be seen to be severely limited.

The West Midlands Study[2]

A study of women in the West Midlands who had recently[3] experienced technical change at work provides data on the ways in which office jobs are changing. The study was exploratory and was intended to gain a direct account of women's perception of the changes brought about in the nature of their jobs, and the context within which they had to carry them out. It was also concerned with women's experience of the implementation process and included issues of consultation, training and management expectations.

Research was based on questionnaires and a smaller number of follow-up interviews. Questionnaires went to trade union district offices with a request to distribute and collect them from workplaces where technical change had recently occurred. Features on the research were also carried by local press and radio, inviting women to whom it was relevant to request a questionnaire. This approach to distribution was an attempt to cover a wide range of workplaces. Interviews were then carried out with a sample of those who completed questionnaires and indicated their willingness to be contacted.

This approach successfully generated a broad-based response; however it is obviously difficult to judge its representativeness. The main biases evident in demographic terms (in comparison with the local female labour market), included an underrepresentation of women with dependents, older women and women from ethnic minorities. In addition, over 85 per cent were full-timers and 70 per cent could be considered as relatively long service employees having been in their current job for over five years.

Overall 282 questionnaires were analyzed,[4] 67.8 per cent of which were completed by women in clerical occupations as defined by the Women and Employment Survey (Martin and Roberts, 1984). This work, which will be referred to as office work below, was carried out in a range of establishments within manufacturing and service industries and those in the public and private sectors. These 191 questionnaires and an associated seventeen interviews will form the basis of the results discussed here.

Of the seventeen office workers interviewed, only two worked for the same employer and then in very different sections and capacities. Five were engaged in secretarial or general office duties. A further two carried out exclusively clerical work. Two worked as library assistants in jobs considered closer to clerical than

Table 6.1: Workers' accounts of the effects of office automation on their jobs

	Column Percentages			
(N = 191)	Level of Interest	Level of Skill	Level of Stress	Level of Sociability
Higher Before Change	14.6	9.9	7.3	16.2
Higher After Change	42.9	57.6	39.8	4.2
No Difference	28.2	18.3	35.0	56.0
No Answer	14.3	14.2	17.9	23.6
Total	100.0	100.0	100.0	100.0

professional occupations. Three worked with Computer Aided Design technology in draughting or information processing occupations. Three workers had some supervisory responsibilities. The final two women interviewed were currently unemployed, one as a direct result of the introduction of new technology. For both their last job had been in an office.

Changes in the Job

When asked how their volume of work had changed, 48 per cent of office workers completing the questionnaire reported an increased workload and only 10 per cent a decreased one. Women were also asked about any changes to the mix of tasks they undertook. Sixty-seven per cent reported that they had taken on some new tasks and only 27 per cent had lost some. Additional information from the interviews suggested that there was often a connection between the task mix changes and the additional workload. Women wanted to, or were expected to, make the most of the new facilities available, but there appeared to be little awareness or acknowledgment of the extra work involved on the part of management. A number of the women interviewed captured this mixture of formal continuity and yet perceived change by saying 'the job is the same but the nature of the job is different'. This concept will be explored further below.

The questionnaire asked women to compare various aspects of their job before and after the introduction of new technology. The results for office workers are summarized in Table 6.1. Clearly such a broadly based survey is in danger of conflating very different experiences. However there did appear to be a fair measure of agreement about the general direction of change.

A large number of clerical workers thus appear to find their job has gained interest since the introduction of new technology; certainly few thought it had lost interest. This positive response was expressed even more strongly in the assessment of skill changes. The assessment of stress levels is less reassuring although it remains a minority of women who say their work conditions have actually deteriorated. Little seems to have happened to levels of sociability of the jobs.

In two open-ended questions women were asked to detail anything, not already mentioned, they had liked or disliked about the technical change. Two-thirds of office workers mentioned some aspect that they liked. Over a quarter of all office workers found some improvement in the nature of the job: it was 'easier to find information' or 'more efficient', or the process of doing the job had changed for the better, for example, 'less typing'. The next most common response (22 per cent of total) was given by workers who felt positively about gaining 'a new skill' related to new technology, or of being given 'a challenge' to tackle something new. A smaller number made reference to the opportunities the new technology provided to produce 'better quality work', or to give a 'better service' to customers. Dislikes primarily related to the process of introduction and will be discussed below.

Findings from two other studies broadly confirm these results. Daniel's (1987) study was based on senior management's and shop stewards' assessment of (male and female) workers' reactions to such changes. For the most part there was agreement that this had been generally positive. The survey further attempted to chart in more detail changes in the jobs of office workers. The findings suggest that according to managers, job interest, levels of skill and range of activity were most likely to increase. Sixty per cent said that job interest had increased, with only 5 per cent saying it had decreased. Fifty-five per cent claimed that skill levels had increased, with only 2 per cent feeling that they had decreased. In the case of changes in the range of activity, 59 per cent felt that this had increased, while only 8 per cent said that it had decreased. Other aspects including level of responsibility, pace of work, control over the way they do their job and level of supervision were considered most likely to have remained the same (Daniel, 1987, Table VII.5, p. 161).

Labour Research (1985) carried out a large scale survey of word processor usage, again mainly based on a third party's assessment of the effects on jobs, rather than direct questions to operators. The focus of this study was on health and safety issues but some discussion of job changes is included. Labour Research received 206 questionnaires from workplaces, with a total of 7,000 VDU screens used by 17,000 male and female workers from all sectors of the economy. Forty-two per cent of responses said that the work now had more interest compared with 16 per cent who said it had less. Significantly this study distinguishes between different types of office computer usage. Subdivision of the results showed that workers who used typesetting equipment were more likely to record a decline in the interest of their work, whereas CAD users were the most positive. However a third of the sample report that jobs are now more stressful compared with only 2 per cent who say that they are less stressful. Again the most negative responses came from typesetters.

Health Problems

A more negative view of the changes was apparent when health problems were discussed. Returning to the West Midlands study, women office workers completing the questionnaire were asked whether particular health problems had increased since new technology had been introduced. Sixty-three per cent of the sample of office workers reported increased eye strain; 50 per cent increased headaches; 43 per cent more tiredness and depression; and 28 per cent increased backaches.

Follow-up interview material suggested that when problems were experienced by the respondent, they were often shared by other women in the workplace doing the same type of work. There was strong recognition of the interaction between various factors in creating health problems. Thus women suggested that health problems were at times caused by lighting, lack of air conditioning, the stress of learning a new job or fears about possible redundancies, as well as being directly attributed to the technology.

Labour Research (1985) found similarly high levels of health problems in their research and they thoroughly review studies which attempt to identify the causes of these problems. There seems to be much evidence that eye strain, while disturbing, causes temporary discomfort rather than lasting damage. Furthermore it appears that many other common health problems can be attributed to system design rather than being an intrinsic problem of the equipment itself. Research on other issues such as reproductive risks is still inconclusive. These, and similar, findings have led to an extensive literature on ergonomic design to minimize problems, and to agreements between management and unions on good working practices.

Despite this, evidence from both the West Midlands study and Labour Research suggests that health problems are still extremely widespread. Interviews with women in the West Midlands study allowed management responses to be examined in more detail. Most women interviewed felt that management had behaved reasonably or even well in response to health concerns. However what this meant in practice was extremely variable. In some cases equipment was moved to a new location, new chairs bought, or break periods enforced. In other cases the response was more superficial: for example, payment for eye tests, or women who were pregnant being allowed to move to other work. In a few cases management had either failed to make any response, or had said that the required changes would be too expensive. One woman said that any mention of problems was met with hostility and felt that this inhibited others from voicing complaints or anxieties.

The Labour Process Scenario

Despite these negative aspects of the change, the research reported above does not suggest the effects on skill are as labour process theorists would expect. There are a number of possible reasons for this. Different possible meanings of the term 'skill' and the changes in work organization that are categorized as 'deskilling' have been extensively discussed in the literature. Furthermore it has been recognized that defining women's work as skilled or unskilled can be particularly problematic, (Phillips and Taylor, 1980). It is therefore important to clarify the sense in which skill is being used in this study.

The comparative measures used in the West Midlands study (e.g. 'Is your job more or less skilled than before?') have the obvious disadvantage of relativism. That is, we are left with no indication of how skilled the job is in comparison with other areas of work; more skilled may still mean virtually no skill at all. There is no clear way around this using self-assessment since each worker has limited experience of other types of work. One can only assume that those who answered this question felt that skill was a meaningful concept to apply to their work. This impression is further reinforced by the significant proportion of women who mentioned skill in the open-ended question on positive aspects of the new technology.

From the questionnaire evidence about the changes in their work and follow-up interviews, it appears most likely that what was meant by skill was being able to do a range of things. Office workers pointed to the reduction in the proportion of tedious repetitive tasks. This meant more time was spent on the less routine elements of the job. Those who used computerized databases found they could more easily access a wider range of information and so could do additional tasks or present information in a new way. Women also identified ability to work on a word processor or computer as a valued new skill in itself, regardless of what it was being used for.

While these uses of the term 'skill' remain task-based (as opposed to person-based or socially sustained; see Cockburn, 1983), they do not necessarily involve issues of control which have been central to labour process debates. Murray (1987) argues that it is precisely the ability to carry out a wide range of tasks, and the associated ability to allocate work time between them, and liaise with others, that characterizes skills in the office. He suggests skill requirements could be better judged on the basis of a comparison between the performance of a new recruit and an experienced worker, than by measures typically used by labour process theorists.

In contrast, Crompton and Jones (1984) do take a labour process approach and distinguish between skill levels on the basis of questions designed to reveal levels of discretion (e.g. 'How much of your work is routine?', 'Are you able to take decisions?'). On this basis they find that researchers and workers ascribe low skill levels to office work. However, the findings of Daniel, discussed above, showed little change in levels of supervision and control over work associated with the introduction of advanced technology (Daniel, 1987).

On the basis of their findings, Crompton and Jones (1984) subscribe to a deskilling analysis of office work. However, as they acknowledge, their approach only gives a 'snapshot' measure. The comparison on which the deskilling analysis is defended is not the job before computerization, but rather some more distant image of clerical work (as described by Lockwood, 1958, and others) outside the experience of the current workforce. Crompton and Jones are therefore effectively seeing deskilling as part of a broad long-term process of rationalization rather than as a direct consequence of the current wave of new technology. Thus their findings are not necessarily in conflict with Daniel's (1987) or those presented from the current research.

It appears that office jobs were 'deskilled' long before the current round of new technology was introduced. It may well be the case therefore that the jobs described are *both* deskilled in the sense meant by labour process writers, and that workers feel they have gained new 'skills' through the introduction of new technology and that their work has become more 'skilled'. The concerns of labour process writers remain important since they relate to an erosion of aspects of work organization which have traditionally been a source of power for the (male) workforce. Equally it is important not to deny the feelings expressed by women about the ways in which their jobs have changed.

The New Opportunities for Women Scenario

Having qualified some of the more negative predictions about the likely effects of introducing new technology into the office, is it possible to endorse the positive view of enhanced rewards and opportunities for access to new types of work? Results from the West Midlands research are not encouraging.

The data on the extent and nature of training provision provided by the questionnaire is relevant here, since it provides some indication of the degree of formal recognition of the way in which jobs have changed. Formal training is also important since the acquisition of recognized qualifications can enhance job mobility. For example, Crompton and Jones (1984) in their study of clerical work point to a trend they call 'credentialism'. By this they mean that opportunities for those who have 'served their time', or who show a particular interest or aptitude for a new area of work and wish to move into it, are becoming increasingly limited as entry and promotion are increasingly only available to those with formal qualifications. Questionnaire findings from the West Midlands study revealed that less than half of women office workers had been given any 'off-the-job' training, and of those who had, 71 per cent had no more than one day's training.

The questionnaire further asked women whether they found their training adequate and how long after training it had taken them to feel confident about using the new equipment. Less than two-thirds of those trained felt the training was adequate. Only 10 per cent said that they had felt confident straight away, and well over 40 per cent said it had taken them more than a month. Zuboff (1989) provides a particularly illuminating account of the nature of the skill changes associated with information technology, which she describes as a shift from action-centred to intellective skills. This makes clear that the treatment of training as a technical task, for example learning what particular keys on the computer do, misses the most important dimensions of both the original job and the ways in which it has changed.

The majority of those office workers interviewed had received training which was highly job specific. Back-up in case of any problems which followed formal training was available to some women, particularly those in large organizations; but the majority had to rely on a manufacturer's manual or discussions with other staff using the equipment. Eight of the women interviewed mentioned that they had subsequently been expected to train other workers in use of the equipment. Only one had received instruction in how to train, or had been given any additional allowances for taking on this task. Her major task is now to supervise YTS trainees in the use of computers. The training course she has attended appeared to be on her own initiative, although her employer did allow her time off to do it.

This practice of getting staff who have already received training to pass on their knowledge to other employees could be a useful way of combining understanding of both the new equipment and organizational procedures. However, judging from the way the process was described by most women in the study (limited support, time or acknowledgment), it seems more designed as a cheap route to getting a minimum level of competence by a necessary number of staff. Given skill shortages and the consequent premium which many firms have to pay to acquire clerical workers with new technology-based skills (Daniel, 1987) the economies seem even greater. Such approaches may appear positive to women receiving training or new technology-based experience from their employers, but their effect (regardless of their intention), seems to be to create low level, company or equipment-specific skills, very limited overall understanding and little basis for job mobility. As such this approach is better seen as illustrative of the widely documented lack of commitment to skills development by British management, with negative consequences for women's opportunities.

It appears that most women were neither given recognition for the changes that had occurred in their jobs, nor given the formal qualifications that might have helped them to gain such recognition elsewhere. More detailed discussions with the seventeen women interviewed reinforced this view. Twelve remained in what their employers classed as 'the same job' after the introduction of new technology. One woman clerical worker volunteered for what was effectively an occupational shift. The technology was an aid to work being done by another occupational group, but management argued that since it was brought in on a trial basis and the most relevant group of workers was understaffed, it should be offered to clerical workers on a temporary basis. She now appears to be doing the job on a permanent basis, although she remains graded and paid as a clerical worker.

Three women had succeeded in getting some promotion through their experience. All had done so by leaving the employer with whom they had first encountered information technology. Their new jobs were supervisory positions within clerical occupational grades. Since this is a well established, if very limited career route, they can hardly be held up as examples of the new opportunities available to women who gain technical experience.

Occupational Restructuring Scenario

The evidence for any occupational restructuring is even more limited. Ninety per cent of women office workers who completed the questionnaire said that there was no official change of job title following technical change and 86 per cent reported no improvements in grading or pay. When contrasted with women's assessment of the significance of the changes, this seems a very low level of formal recognition of the extent of job change.

In the only example of radical change within our interview sample, the clerical worker was made redundant. The change involved an expansion of professionals' jobs to encompass her previously lower graded work. This could be seen as a reintegration of a previous division of labour, but not one that appears to benefit clerical workers!

Part of the reason for this finding seems to be that technical change is currently being implemented in a far less radical way than the 'electronic office' visions would have us believe. Despite the potential for restructuring communications and information processing, most office technology is being used for little more than enhancing the productivity of routine tasks and, to a limited extent, the diversity of products. The current range of tasks undertaken by most office workers is not being fundamentally restructured, although the ways in which they are doing them may be. It is perhaps not surprising that such changes do not open up many new possibilities for women.

Trends described by some writers could actually entrench women's currently subordinate position. Knights and Sturdy (1987) examine changes in the insurance industry associated with the introduction of information technology, and argue that there has been a massive increase in routine work. They suggest that this has led to a polarization of skills whereby some jobs are no longer seen as having any place on career paths. These jobs are increasingly separated, both physically and in terms of knowledge requirements, from higher grade work, thus making movement from one job to another very unlikely. Rajan (1985) documents more broadly, similar trends within the finance sector. He suggests that there is likely to be an increasing divergence between clerical and professional jobs, reflected at recruitment level through differential entry requirements.

The potential for such developments to be linked to discriminatory recruitment strategies, has been recognized by the Equal Opportunities Commission in their investigations of various financial institutions. The tendency for men to be promoted from clerical and low grade administrative jobs in preference to women doing similar tasks, is explicitly discussed in studies by Crompton and Jones (1984) and by Davies and Rosser (1986).

Such findings are often explained by pointing to factors outside of the workplace. It is argued that girls are socialized and educated differently from boys leading to different interests and competencies. Women's ability to commit themselves to training or to seek responsible jobs is restricted by domestic demands. Women's career paths are disrupted by time off for childbirth and childrearing, and the tendency to prioritize husbands' careers (for example in decisions about household moves). Thus, it is suggested, technical and organizational factors determine the characteristics of the jobs which are available, while broader social factors determine the distribution of men and women within them.

Attention to the process of technical change suggests that factors perpetuating the sexual division of labour cannot be separated out as simply as this analysis suggests. The tendency to contain changes in the range of tasks, skills and workload levels within a continuity of job title and grade, is likely to restrict any re-evaluation of either the job holder or the position of the job within the occupational hierarchy. Highly job-specific and limited training, the tendency for women themselves to train other workers and for this not to lead to formal qualifications, all militate against the type of understanding or prerequisites to expand one's job or move to another more challenging one.

While one might expect technical change to disrupt the current sexual division of labour in the office, it appears that changes are only occurring in *the way a job is done* rather than in *what is being done*. This seems to be what was meant by women in the West Midlands study when they said 'the job's the same but the nature of the job is different'. I have argued elsewhere, in relation to manufacturing jobs (Liff, 1986), that this type of change is unlikely to lead to a questioning of occupational sex-typing. Indeed it would appear that the structures which currently contribute to women's subordination within the workforce are actively reproduced by this form of technical change.

Overview of the Process of Change

The picture that emerges for women office workers is mixed. For many there is some improvement to the quality of their work experience, but they remain trapped within segregated, low grade work. Examination of current trends suggest that this is unlikely to change as a fortuitous side effect of current forms of technical change. The reasons for the lack of straightforward evidence for the deterioration of work experience predicted by the labour process approach, has already been discussed. The lack of formal training and qualifications has been advanced as one important reason why women are not able to use their experience with new technology as an effective lever in gaining access to better jobs. The limited form of office automation currently being introduced was seen to contribute to the lack of challenge to occupational boundaries and sex-typing.

Opportunities for Influencing System Design and Implementation

In contrast to the generally positive assessment of the changed nature of the job, a large number of office workers in the West Midlands study were dissatisfied with the way change was managed. Forty-seven per cent of office workers detailed something they disliked about the change. The biggest category was those who referred to the process of introduction. This usually related to problems with the type of or lack of training given, lack of consultation, or to management's expectations about levels of performance. The latter included being expected to use and maintain both the manual and computer systems at the same time, and to take on extra tasks 'because they think the machine does all the work now'.

Consultation and Participation

A willingness to consult with the workforce over technical change or more positively to encourage participation in decision making, could be an indicator of women's ability to influence the process of change. Such an approach is also recommended as sound business practice, since it is thought to smooth the adoption process. Mumford and Banks wrote in 1967 that the introduction of computers into the office was causing anxiety and antagonism as a result of

> ... management's unawareness of the social implications of the change they were introducing .... [their] defective communication and consultation policies.... [and] the pressures of the change period. (p. 15)

These problems could only be resolved, they believed, if management 'take into account technical, organisational and human relations factors' (p. 17). They went on to suggest a way of handling technical change based on this analysis. Since then a vast literature has developed on this subject.

Despite this, current studies continue to show little evidence of this approach to the management of change. Daniel (1987) expressed surprise at the low level of consultation which occurred in the majority of companies adopting new technology. For manual workers, consultation tended to be restricted to those circumstances where either this approach was forced upon management by established agreements with trade unions, or where there was strong, early evidence that conflict was likely. In contrast, for office workers, consultation was much less likely to be dependent on such pressures on management. Formal consultation with trade union representatives was lower than for manual workers, even when the sample was restricted to organizations within which unions were officially recognized. However, informal consultation was far more common. Managers reported that they had discussed the changes with individual office workers in 82 per cent of cases, but with individual manual workers in only 62 per cent of cases.

Daniel regards this degree of informality as positive evidence of a lack of social distance between management and office workers. However there are less positive interpretations, such as a failure to take women workers' contributions seriously. For example, an EEC study (Chalude, 1984) which analyzed the introduction of office automation in a small number of Belgian and French firms, found that consultation was minimal and management gave patronizing assessments of the women's interest, or ability to make a real contribution. Huggett (1988) describes a British case-study which makes similar points. These studies suggest it would be wise to treat with some caution managers' reports of the extent to which they consult with office workers. The EEC study also questions the view that this is just a 'British problem' that can be resolved as we move closer to European approaches to labour management through the Social Charter.

Women interviewed for the West Midlands study were asked how they had first heard about the change, and whether they had had any opportunity to express their views. In only two cases did women have any say about whether the change itself was necessary. Interestingly, these were situations where women 'bullied' management to acquire equipment which they felt they needed. In neither case did the involvement go beyond making a general case for adoption. In no case did the women interviewed have any input to the decision about what specific technology to adopt. In one case the woman in her role as a senior secretary was taken to a VDU demonstration. For the other women, the first real picture they had of the new technology was when it arrived. Periods of notice given before the appearance of new technology varied from a few months to none at all.

Mumford (1983), one of the best known proponents of participative approaches to system design, gives a more positive account of the way office workers can become involved in technical change. She describes her role as a facilitator of a group of secretaries at ICI who developed recommendations for improved word processor usage. The group gained considerable expertise and confidence to express their ideas about the ways in which they wanted work to be organized. Her account thus provides valuable evidence against the view that 'semi-skilled' workers are incapable of making a contribution to system design. It also provides evidence that word processors can be used in ways that satisfy workers and are seen as efficient by managers. More problematic is the extent to which this can form a general model for office redesign. Issues about the resolution of conflict between different interest groups are not adequately dealt with. Also, the commitment of managerial and secretarial time to the project was extensive and it seems unlikely that many organizations would feel able to devote this level of resources to the task. There is at least a strong implication that the findings of the study would be used even in this organization as the basis of more extensive office redesign. Thus this highly participative exercise would seem likely to apply to a very limited number of staff.

Trade Unions

Where management are doing little either to increase women's involvement in the process of change or take account of their concerns, can trade unions represent their interests? Because of the distribution method used, the West Midlands study contained a large proportion of trade union members. Of the seventeen office workers interviewed, fourteen were currently members of a trade union and one other woman had been a member but had left prior to the introduction of new technology. Where women in the interview sample were represented by trade unions, they were asked whether the union had had any involvement in the decision-making process. Most were unsure about what had occurred between management and the trade union but were clear that *their* views had not been solicited by the union. When asked more specific questions about whether the union had had any say in choice of system: timing, work organization or training, all but one said that they knew it had not, or thought it highly unlikely. At best, therefore, it seemed the union had been told about the decision at a similar time to the workforce. In one case where a trial was involved, the union did negotiate rights to return to the original job if requested. As expected, other comments focused on health and safety related agreements. A similar level of lack of knowledge on the part of office workers of the position being taken by their union over new technology is reported by Storey (1986).

The broader literature on trade union bargaining and new technology suggests that such situations are not uncommon. In an analysis of new technology agreements, Williams and Steward (1985) suggest that trade unions are rarely able to gain rights to be consulted at an early stage in the decision to implement new technology, and may have difficulty in making use of the information they do have access to. Lane (1986) makes similar points in an analysis of unions' ability to participate more generally in the planning process.

Many writers have pointed to the specific problems faced by women within trade unions. Because of occupational segregation, trade union officials (predominantly men) are likely to have limited knowledge about jobs being done by women. This again contributes to the difficulties union negotiators experience in assessing change. It may also lead to conflict of interest where occupational change is occurring. The one case of redundancy in the interview sample partly illustrates such a case. It also illustrates the ways in which the interests and concerns of someone outside of a powerful occupational group can be completely ignored.

The woman who was made redundant had been aware of the possibility of change for a long time, since the union which represented another occupational group was engaged in negotiations over the terms on which the latter would accept the new technology. While their opposition continued her job was safe. She was neither informed officially of the possibility of change by management, nor involved by anyone in discussions about her position. Hearing that the union had completed negotiations to their satisfaction, she drew her own conclusions. Soon after, she was informed that she was to be made compulsorily redundant with minimum notice and payment. The woman concerned had left a trade union over an earlier issue where she felt her interests had been inadequately represented. She said she now regretted this, as she thought she would have still been made redundant but the compensation might have been better!

Traditional forms of trade union bargaining based on compromise between polarised positions and concentration on a narrow range of issues may exacerbate these problems. For example, Williams and Pierce (1984) report a case where management and trade unions negotiated a compromise on the 'refresh rate' for VDU screens, which was half way between that originally proposed by management and the trade union. This was written into a technology agreement and adopted in a number of other situations. However equipment was not made to these specifications so the agreements were inoperable. Discussions with NALGO officials in relation to the West Midlands research highlighted another problem. Officials had included a number of agreements which stated that workers should not operate VDUs for more than four hours per day. But without negotiations covering job redesign, there was a danger that workers would not have sufficient non-VDU related tasks to make this viable. Women without formal representation who attempt to negotiate on an individual basis with their managers, are likely to find all these problems magnified.

Equal Opportunities Policies

Equal opportunities policies currently focus on removing barriers to entry to occupations. They do this by stressing fair competition for pre-defined jobs primarily via formalized assessment procedures (Liff, 1989). While such strategies have allowed some women to enter traditionally male jobs, the broad pattern of sex-typing remains. Cockburn (1985), Game and Pringle (1984) and others have argued that the continuing existence of occupational sex-typing plays a fundamental role in perpetuating gender identity. Cockburn (1987) argues further, that jobs and the skills which are seen as necessary to carry them out, often incorporate gender stereotypes. This not only affects selectors' perceptions of 'suitability', but can also lead individuals to feel that their own sense of gender identity is threatened, if they attempt to cross traditional occupational divides.

For these reasons Webb and Liff (1988) argue that equal opportunities policies should place more emphasis on trying to improve women's opportunities within the jobs they currently occupy. There is a need for policies requiring employers to rethink job requirements more deeply, and restructure jobs across existing occupational boundaries. Technical change could well provide a focus for this type of re-evaluation, although a study by Wilson (1987) suggests that such connections are not easily made by managers. It seems more likely that many initiatives will come from groups or organizations with an explicit commitment to equal opportunities.

Work described elsewhere in this volume by Hales and O'Hara (Chapter 9) and by Green, Owen and Pain (Chapter 8) derive from this perspective. As long as such approaches are treated as marginal by management there will be little likelihood of women benefiting in more than a limited way from technical change in the office.

Notes

1 An earlier version of this paper appeared as 'Clerical Workers and Information Technology: Gender Relations and Occupational Change' published in *New Technology, Work and Employment*, **5**, 1, 1990 and is reproduced here with the permission of Basil Blackwell.

2 The research described was funded in 1986 by a grant from the West Midlands County Council to the West Midlands Women and Technology Group. In particular

I would like to acknowledge the help of Tina Mackay and Philippa Williams with data analysis.

3 Research suggests change can better be understood as a process rather than as a single event occurring at one particular point in time. For this reason we did not seek to define 'recent' in a rigid way. However discussions suggest that most women were talking about changes which had occurred over the previous year.

4 Response rates varied enormously depending on distribution routes. Where we were able to place questionnaires directly in workplaces, interest and level of response were high. In other cases batches of questionnaires were sent to trade union offices on the basis of an initial contact. Follow-up contact suggested that in many cases these had become stuck in 'the system' and never reached appropriate workers.

References

ARMOUR, H. (1986) *New Technology in the Office Environment*, Aldershot, Gower.

BARKER, J. and DOWNING, H. (1980) 'Word Processing and the Transformation of Patriarchal Relations of Control in the Office', *Capital and Class*, **10**.

BEVAN, S.M. (1987) 'New Office Technology and the Changing Role of Secretaries', in DAVIDSON, M.J. and COOPER, C.L. (Eds) *Women and Information Technology*, London, John Wiley & Sons.

CHALUDE, M. (1984) *Office Automation and Work for Women*, Brussels, Commission of the European Communities.

COCKBURN, C. (1983) *Brothers: Male Dominance and Technological Change*, London, Pluto Press.

COCKBURN, C. (1985) *Machinery of Dominance: Women, Men and Technical Know-How*, London, Pluto Press.

COCKBURN, C. (1987) *Two-Track Training*, London, Macmillan.

CROMPTON, R. and JONES, G. (1984) *White-Collar Proletariat: Deskilling and Gender in Clerical Work*, London, Macmillan.

DANIEL, W. (1987) *Workplace Industrial Relations and Technical Change*, London, Frances Pinter.

DAVIES, C. and ROSSER, J. (1986) 'Gendered Jobs in the Health Service: A problem for labour process analysis', in KNIGHTS, D. and WILLMOTT, H. (Eds) *Gender and the Labour Process*, Aldershot, Gower.

DEX, S. (1987) *Women's Occupational Mobility*, London, Macmillan.

GAME, A. and PRINGLE, R. (1984) *Gender at Work*, London, Pluto.

HMSO (1986) *New Earnings Survey*, London, HMSO.

HUGGETT, C. (1988) *Participation in Practice*, Watford, EITB.

KANTER, R.M. (1977) *Men and Women of the Corporation*, New York, Basic Books.

KNIGHTS, D. and STURDY, A. (1987) 'Women's Work in Insurance — Information Technology and the Reproduction of Gendered Segregation', in DAVIDSON, M.J. and COOPER, C.L. (Eds) *Women and Information Technology*, London, John Wiley & Sons.

LABOUR RESEARCH DEPARTMENT (1985) *VDUs, Health and Jobs*, London, LRD Publications.

LANE, T. (1986) 'Economic Democracy: Are Trade Unions Equipped?', *Industrial Relations Journal*, **17**, 4.

LIFF, S. (1986) 'Technical Change and Occupational Sex-typing', in KNIGHTS, D. and WILLMOTT, H. (Eds) *Gender and the Labour Process*, Aldershot, Gower.

LIFF, S. (1987) 'Gender Relations in the Construction of Jobs', in MCNEIL, M. (Ed.) *Gender and Expertise*, London, Free Association Books.

LIFF, S. (1989) 'Assessing Equal Opportunities Policies', *Personnel Review*, **18**, 1.
LOCKWOOD, D. (1958) *The Blackcoated Worker*, London, George Allen & Unwin.
MANNELL, M. (1987) 'A Career Path for Secretaries', *Training Officer*, February, p. 45.
MARTIN, J. and ROBERTS, C. (1984) *Women and Employment Survey*, Department of Employment and OPCS, London, HMSO.
MUMFORD, E. and BANKS, O. (1967) *The Computer and the Clerk*, London, Routledge and Kegan Paul.
MUMFORD, E. (1983) *Designing Secretaries*, Manchester, Manchester Business School.
MURRAY, F. (1987) *Reconsidering Clerical Skills and Computerisation in UK Retail Banking*, Working Paper, Departments of Computer Studies and Applied Social Studies, Sheffield City Polytechnic.
NATIONAL ECONOMIC DEVELOPMENT COMMITTEE (NEDC) (1983) *The Impact of Advanced Information Systems*, London, NEDO.
PHILLIPS, A. and TAYLOR, B. (1980) 'Sex and Skill: Notes Towards a Feminist Economics', *Feminist Review*, **6**.
PRINGLE, R. (1989a) 'Bureaucracy, Rationality and Sexuality: The Case of Secretaries' in HEARN, J. *et al.* (Eds) *The Sexuality of Organization*, London, Sage.
PRINGLE, R. (1989b) *Secretaries Talk*, London, Verso.
RAJAN, A. (1985) 'New Technology and Jobs: The Counter Arguments', *Personnel Management*, July.
STOREY, J. (1986) 'The Phoney War? New Office Technology: Organisation and Control' in KNIGHTS, D. and WILLMOTT, H. (Eds) *Managing the Labour Process*, Aldershot, Gower.
WEBB, J. and LIFF, S. (1988) 'Play the White Man: The Social Construction of Fairness and Competition in Equal Opportunities Policies', *Sociological Review*, **36**, 3.
WILLIAMS, R. and PIERCE, B. (1984) 'Design, New Technology and Trade Unions' in CROSS, N. (Ed.) *Design and Society*, London, Design Council.
WILLIAMS, R. and STEWARD, F. (1985) 'New Technology Agreements: An assessment', *Industrial Relations Journal*, **16**, 3.
WILSON, F. (1987) 'Women, Office Technology and Equal Opportunities — The Role of Trade Unions' in DAVIDSON, M.J. and COOPER, C.L. (Eds) *Women and Information Technology*, London, John Wiley & Sons.
ZUBOFF, S. (1989) *In the Age of the Smart Machine*, London, Heinemann.

Chapter 7

From the Word Processor to the Micro: Gender Issues in the Development of Information Technology in the Office

Juliet Webster

Introduction

It has often been assumed by those writing about the emergence and effects of information technologies that the latter are autonomous, developing according to some inner logic and driven forward by their own momentum. In this conception, technological development proceeds along some predetermined and inevitable path, in a single direction. Its evolution is unproblematic and reflects the 'best way' of addressing practical and technical issues. Moreover, since a technology consists of objective improvements upon previous techniques and travels along a single path of development, it has a single set of consequences for patterns and processes of work and organizations.

This perspective is problematic. It prioritizes the technical component of technological change, and it suggests a causal, linear relationship between technical and organizational change. In so doing, it fails to grasp the non-technical factors which contribute to the development and form of technologies: the economic, political, social and organizational artefacts and relationships which condition the path of technological development and which give technologies their ever-changing form. It also understates vital elements in the relationship between the technical and the social. Since only the influence of the former upon the latter is recognized, the ways in which technologies and social structures mutually interact is missed. In short, the complex nature of the correspondence between social arrangements and technologies is completely overlooked.

In the case of information technologies (ITs) in the office, this kind of approach has meant that many analyses of their emergence and impact have treated IT systems as though they had an inevitable path of development, governed solely by macro-level imperatives to increasingly subordinate office work to automation.[1] The emergence of IT systems has therefore been largely considered in isolation from the particular office contexts within which they are applied, and without recognition of the particular groups of workers at whom they are targeted. Moreover, and consequently, their effects have often simply been 'read off' from these macro-level imperatives, and sweeping generalizations (of both an optimistic and a pessimistic character) have been made about the nature of work in the automated office and about the people who perform this work. These generalizations, effectively postulated trajectories of work organization development, have often treated as uniform the different office labour processes to which technologies are applied, and have also failed to distinguish between the effects of technologies and changes in labour process upon male office workers on the one hand and female office workers on the other. It has often been assumed that technical change will have a uniform set of consequences for all office workers; the effects on male, skilled workers will be the effects on all.[2]

The introduction of IT systems into the sphere of office work has been a much more complex affair. Trajectories of equipment and organizational development have not followed the straightforward, linear route expected of them.[3] Nor has their relationship been one of technological determinism, with the equipment simply having social and organizational 'impacts'. Instead, both office automation and the automated office have mutually conditioned one another and thus have taken a highly varied form (Thompson, 1983). The logic behind IT equipment design has not simply reflected technical or even economic rationality, but has also reflected the gendered nature of the work for which it has been designed. Office technologies have also been fluid in form, undergoing fundamental changes over the last couple of decades. Moreover, once established, these systems have been configured in various ways which reflect local office contexts and requirements as much as hardware and software exigencies, and which have had crucial impacts on the design of subsequent IT generations.[4] The implementation of information technology in the office, then, has involved the fusion of technical and social/political elements such as work organization, managerial practices, gender relations and expertise which are in mutual interaction.[5]

This chapter provides an empirical view of the evolution and impact of a key component of office automation: the word processor. It considers the trajectory of development of this technology, showing how the design and evolution of the equipment has diverged from its original path. It demonstrates that social structures, and class, gender and interpersonal relations within workplaces may influence and indeed confound the direction of technological change. The chapter then examines the impact of word processing systems upon the area of office work most fundamentally and visibly affected, namely, women's secretarial and typing work. It looks at the expectations that were held about the impact of word processing upon work organization and skills, and compares these with the actual experience of automated word processing ten years on. It highlights the various factors which have shaped the implementation experience of these systems and which are critical for an understanding of the changes in technology and in the organization of office labour processes which have taken place over the last decade. The interplay of class and gender relations in the office and in society at large emerges as central to this understanding. The chapter argues that accounts which postulate simple, unilinear trajectories of development, both technological and organizational, fail to provide full insights into the changes taking place in the office in the late twentieth century.

The Evolution of the Word Processor

Word processing is the outcome of a number of different innovations taking place in different fields of electronic text production and reproduction. Its origins are threefold and distinctive: data processing, programming, and office equipment.

In the early business systems of the 1950s and 1960s, computing was carried out on large-scale mainframe machines which were used entirely for data processing applications. There were both software and hardware constraints to extending the domain of these computing giants to cover other business applications involving the processing of text. First, screen editing was not possible at that time; all codes and data were line-edited. This was a tedious process, which militated against the wholesale inclusion of word processing as an application on mainframe machines. Moreover, word processing was highly resource-hungry, involving heavy usage of disk and memory space to do the writing and rewriting, saving and re-saving which is typically involved. Word processing would thus have to compete with the more established data processing applications for meagre computing resources. We can speculate that perhaps one reason why data processing was favoured over text processing applications on mainframes was that text processing was seen as a 'soft' application, not a proper utilization of computing resources. There was certainly an ethos that computers were a 'hard' mathematical technology, built for calculating rather than composing. Most early industrial applications of computers involved their calculation, algorithmic and file-handling functions, financial and accounting applications, for example. Initially, these took precedence over activities which were less quantitative in nature, the processing of text, for instance. This latter was, of course, also defined as 'women's work', and as such was performed using relatively cheap labour-power, never a priority for managements' automation programmes. This may also explain why word processing was not immediately a target for automation. There was therefore an identifiable technical logic — and arguably also a social imperative — in the emergence of text processing packages such as Textjab, as the solution to these problems. Files of text were created and stored unformatted, and were then formatted by being run through such a package.

Word processing also evolved in the computer programming sphere. The early programmers who were writing software in machine code needed some kind of tool for editing this code. A package was developed for this purpose, and thus, Word-Star was born. It was not originally meant for office use at all, or for word processing proper. In fact, it was only with the take-off of micro-computing in the 1970s, as a result of the development of spreadsheet systems like Visicalc, that Word-Star became predominantly marketed and used as a full word processing system with wider applications than simply editing code.

The third line of descent of word processing is not computing at all, but office equipment. From the mid-50s onwards, office technologies began to undergo a series of enhancements to incorporate increasingly sophisticated facilities for the production of documents, a process which was led by the then IBM Electric Typewriter Division.[6] In 1964 the Magnetic Tape/Selectric Typewriter (MT/ST) was announced; this was a device which was essentially a typewriter with tape storage and a slight delay between keying and printing, which allowed the typist to create rough drafts very rapidly, backspacing and correcting errors as they were made. IBM labelled the concept behind this technology 'power typing', and from the late 1960s onwards it introduced a range of products in its magnetic media line. Some of these incorporated the storage of basic programs for formatting text, as well as the text itself; facilities included justification, centering of text and positioning of text flush to left-hand or right-hand margins. Other manufacturers experimented with cathode ray tube display workstations, and in 1972, the development of daisywheel printer technology provided letter quality printing at high speeds and thus increased the popularity of these putative word processing devices. In 1974, the floppy disk, already in general use for computing applications, was imported to word processing as the storage medium. These combined innovations brought into being the dedicated word processor technology which diffused widely into the offices of the 1970s and early 1980s.

The Social Shaping of Word Processing

Why, in its first generation, did this latter word processing technology emerge as dominant? Why did this particular technological form become prevalent over others? The question cannot be understood simply in terms of the 'logic' of straightforward technical advance or efficiency, a logic which tends to be taken for granted in technologically determinist accounts. This does not explain why, initially, word processing took the form that it did. The office equipment-based form of word processing in fact utilized one of the least efficient and most laborious character recording schemes that was developed at the time; even by the standards of the 1960s, this specification was extremely 'low tech'. In order to understand the logic of the development of word processing, we have to look beyond the technical inputs into these devices, to the social context of their development, to the work processes which they were designed to address, to the characteristics of their users. Clearly, a technological application will critically affect the design of that technology; in the case of word processing, this means that its application to particular office tasks and to a certain group of users, in this case women, has had a crucial influence upon the form that it has taken.

The emergence and rise to dominance of a particular word processing design out of the contest between the various innovations, has been driven as much by the work processes involved in text production and by the gender of those who do this work, as it has been by purely technical exigencies. Text production has of course traditionally been performed on typewriters, by women. The most marketable and easily accepted design was likely to be one that was familiar to, and usable by, women office workers. The early word processors therefore utilized the typewriter design, and extended typewriter technologies. They were specialized machines designed for use by trained touch-typists (David, 1985), with QWERTY keyboards and an extra pad of specialized command keys labelled with their functions ('Insert', 'Delete', and 'Save' keys combined with 'Word', 'Line', Paragraph' and 'Document' keys which could, for example, save a document at the press of two dedicated keys). Many word processors also incorporated a screen design which was intended to represent the sheet of paper. This form of word processing, then, emulated the typewriter, in order to make the new technology as accessible to the typist as possible. In terms of its operation and design, it was a relatively minor departure from the old technology, for it was one that incorporated office activities done traditionally by women into its construction and was designed for use specifically by women office workers. It brought electronic technologies to the typewriting task, rather than taking text production technologies to the computing activity. This was to come later, as we shall see.

So the first generation of word processing, like the activity which it automated, was strongly gendered, strongly shaped by those who performed this work. Delgado (1979) points out that when typewriting technology emerged at the end of the nineteenth century, the women who operated them bore the same name as the machines — 'typewriters'. Thus the technology rapidly became synonymous with those who performed the work, and the developing strong association of women with this function ensured that typewriting became a highly gender-specific activity. It was in the context of this gender-specificity of a work process that word processing technology emerged, and the term 'word processor' has also sometimes been used to refer to the incumbents of such office jobs as well as to the devices upon which they work. Because this kind of office labour was historically almost exclusively female, this gendered division of labour became incorporated in the technology just as much as more tangible elements of hardware. In fact, in its own way, the traditional gendered division of office labour has crucially determined the initial direction taken by the technology. This in its turn meant that the first generation of word processing, designed and marketed for use by women office workers, was initially targeted at and used by them, rather than being applied to other white-collar tasks involving the handling of text. In this way the association of women with text production, albeit now automated, was perpetuated. And it was the continuation within word processing of older techniques which themselves embodied a gendered division of labour, coupled with its implementation into offices with gendered but varied patterns of work organization, which influenced the character of the first generation of automated offices.

The Spread of Word Processing and the 'Office of the Future'

When word processing began to diffuse into offices in the late 1970s, gloomy predictions were made about its likely impact upon women, their work and their skills. It was commonly argued that this technology would bring the regimentation of the assembly line into the office, divesting women of all the skills and abilities which they used in the course of their work.[7] Word processing, it was argued, was a mechanism for the introduction of Taylorism into the office and as such, would be associated with the degradation, deskilling and intensification of office work. This suggested that there would be an identifiable trajectory of development in the organization of work in the automated office.

Since these forecasts were made, however, the experience of information technology in the office has proved quite different. The 'impact' of word processing technology (if technology alone can be said to have independent impacts) has been found to be much less straightforward, and much more closely tied up with the context of the work organization into which it is introduced. In the same way as the technological trajectory of development for word processing has not been unidirectional, following a single, expected course, so too there has been no simple trajectory of work organization associated with word processing. Instead we have seen the continuation of different patterns of work organization which have been shaped less by purely technical influences, than by long-term management practices in particular firms; strategies for the control of women's work, national economic and local labour market conditions, and in this context, corporate objectives in introducing new techniques. Empirical research has shown the strong continuities between the organization of work in non-automated offices and in offices where information technologies have been introduced, see, for example, Silverstone and Towler (1983) and Webster (1990b).

Instead of all office workers being reduced to uniform deskilled automatons, the differential between secretarial and typing workers, and the ways in which their respective jobs are organized, persist. As Arnold *et al.* rightly note:

> The distinction between secretarial and typing work is crucial to analysis of the likely effects of WP in offices . . . female typists — essentially machine operators — have existed as long as the machines themselves. As specialist machine operators, they can be managed like factory shop floor workers; this may not be true for secretaries, who perform a range of other duties in addition to typing. (1982, 60)

Indeed, secretarial workers continue to do a variety of tasks, with considerable discretion and technical skill, working very much at their own pace and according to their own pressures. Typists, whose work has always been relatively routinized, continue to perform a much smaller range of more fragmented tasks than secretaries have, with much less discretion and more external direction (often from a typing pool supervisor). Certain kinds of employers, financial institutions in particular, have very long-standing systems of work rationalization which predate, and actually form the basis for, the introduction of office automation systems. For example, the building societies of the 1970s already had highly automated, highly structured labour processes. They regarded themselves as having an organizational system of word processing in place. Correspondence was routinized and simplified: pads of pre-printed letters with tick boxes were used, and there were strict divisions between clerical staff and the typing pool for handling different categories of work. One building society Organization and Methods Manager simply saw the company's word processing system as 'an accumulation of what had gone on before'. By this he meant that the organizational rationalization of work had taken place over a long period and considerably before the change in technology which we now recognize as word processing.

So there are significant continuities in the conduct of office work before and after the introduction of word processor technology. Secretarial workers still have a range of activities, while the regimentation of the typing pool has carried over from the days before office automation. However, many office workers of all categories continue to bring considerable expertise and competence, of both a technical and broader organizational nature, to bear in the course of their work, and these have by no means been eradicated by the introduction of new technologies into the office. It is the case though, that these competences often go unrecognized by the observers of office labour processes, including managements who are unfamiliar with the detail of these labour processes. It is therefore very easy for women's abilities to be downgraded and denigrated, not just by their superiors, but by academic commentators anxious to demonstrate the degradation to which all automated work is subject.

The competences that women office workers bring to bear in their work vary from the organizational skills of the secretary who ensures the smooth running of the office and deals with the office politics which are never explicit, to the typist or word processor operator who, despite the most routine of jobs, still knows how to get the best from her machine and the workings of her particular office environment. The secretary who effectively runs the office without, predictably, any recognition, is a familiar phenomenon. Certainly, this may consist of doing a great number of domestic tasks which typify the work of 'the office wife' (Benet, 1972). However, through the administrative side of their work — dealing with phone callers, arranging internal meetings, processing salaries and pensions, and acting as gatekeeper for their bosses — they often develop a strong familiarity with and sometimes influence over, the running of the firm. Women usually have this influence in the 'private domain' (Stacey, 1960), but secretaries can be influential in the 'private-within-the-public' domain of the office. The secretaries in my study of word processing in West Yorkshire were constantly going round the office to arrange meetings and appointments with other managers; one preferred to do this personally because of the smallness of the company, which meant that she quickly became drawn into office politics and negotiation processes between different members of management (Webster, 1990b). She became a repository of organizational knowledge, and soon people stopped in her office, *en route* to visit her boss, whose office was next door to hers, when they wanted to discuss certain aspects of company business or to exchange information. This organizational expertise and influence has been well described by Vinnicombe (1980) in her study of secretarial work:

> At the top of organisations, where secretaries traditionally operate in the one-to-one working relationship with their managers, secretaries carry out a variety of administrative tasks. These tasks all tend to stem from the secretary's gatekeeper position in her boss's communication network. Theoretically, this position gives the secretary almost complete control over the boss's communications. It also means that she has the opportunity to wield a great deal of influence. The extent to which she can influence matters . . . is also related to her personality and the number of years she has worked for the organisation. The last point is important and frequently underestimated. Many top secretaries have well-developed personal contacts throughout the organisation and have an extensive knowledge of the organisation's activities, and sometimes even its secrets.

Even the most routine of office workers, without the level of autonomy and discretion that characterizes secretarial or supervisory work, commonly know far more than their superiors about the organization of the office, the labour process and, within this, the operation of technologies. Superiors, without understanding what this work entails, commonly undervalue it. One woman who worked in a word processing pool in West Yorkshire commented:

> Terry thinks he knows, but he hasn't a clue what goes on. He isn't going to hear this, is he? They think it is easy [working the word processor]. They think you just press a button and that is it. It takes a lot more doing than they think. It still has to be typed out the same. They just think you press a button and it types it itself. Terry even thinks it types on its own.

Fed up with being constantly undervalued and oversupervised, these typists decided to embark on a campaign to illustrate how difficult their work was made by poor quality audio dictation, and thus to demonstrate what unrecognized competencies they actually required:

> If they said 'er', we put 'er' in. Or they had a bad habit of saying, 'Dear Sir, We have none of your pistons in stock, number RA247 . . .' And then they say, 'Oh, no. We have some of your pistons . . .' That is no good because you have already typed it if you are on a typewriter. You are only about one word behind all the time. So what we did was, we typed exactly what they said. You just can't correct it when you are on a typewriter; you have got to pull the whole thing out and start afresh when they go and say 'Oh no' and then dictate another sentence. So some of them came back and were furious about this and said, 'This is absolute rubbish'. We said, 'Well, that is what you dictated, so we typed it'. They wouldn't admit that they had done it . . . And they do. They are really bad. They say 'Oh no, typist', so we typed 'Oh no, typist'. They didn't like that at all.

Both management and radical critics of the 'office wifely' aspects of secretarial work have contributed to the denigration of the skill or knowledge involved in some office jobs. Management vest intelligence in themselves or in machines, usually because, as the quotation above indicates, they have little understanding of the actual processes involving in performing certain office tasks, or of the competences required. Radical critics, too, over-influenced by both managerial ideology and Bravermanian arguments about the lack of skill required to perform tasks in the modern office, attribute little if any skilled activity to the office worker. While it is important not to romanticize the abilities required to operate the word processing technologies of today's offices or to perform general secretarial tasks, it is also the case that women office workers themselves clearly vest their activities with some levels of skill and expertise. Secretaries and typists commonly express frustration with attitudes which rob them of any recognition of their capabilities.

Part of the problem lies in the fact that both management and labour process writers have tended to define 'skill' in very conventional terms, in relation to the characteristics of male skilled labour. Women's labour, no matter how much technical dexterity, mental expertise or training it requires, is usually defined as inferior simply because it is women's labour (Phillips and Taylor, 1980; Wajcman, 1991). Women's skills come to be defined as non-technical and so undervalued, because, as Cynthia Cockburn (1985) has argued, know-how and technical competence are resources that confer potential or actual power and this has been central to the sexual and class politics of technological work. As Rosemary Pringle (1988) points out, in fact the skills and interpersonal relations and even power involved in women's office work, particularly secretarial work, make it difficult to define their status and position in the office. Certainly they cannot be regarded as simply unskilled or lacking in expertise (Phillips and Taylor, 1980).

The unrecognized knowledge wielded by women office workers may give them unexpected, but not always positive, elements of control over the functioning of the office. A typing pool supervisor in my West Yorkshire study spent most of her time trying to foster the illusion that the pool was efficiently churning out its work. Unfortunately, the typists were so poorly trained and so ignorant of the overall workings of the organization that their standard of work was very low, and as a result, the users of the pool were losing confidence in it and doing as much of their own typing as they could. The supervisor's response to this vicious circle was to give the typists only the most simple and undemanding tasks, and to send the important or complex work out to a typing bureau. At the same time, the typists had to be seen to be looking busy, so she would make them duplicate each other's batches of letters (without telling them that this was happening). Moreover, the word processor which the company had bought stood idle in the corner of the typing pool; its use could not be readily incorporated into this subversive system of work organization.

Despite the assumptions frequently made by their male superiors that women office workers have only trivial expertise, and the assumptions of many academics that information technology would divest them of any expertise, secretaries and typists do continue to exercise both organizational and technical competence, as we have seen. In fact, particularly in relation to office technologies, women possess much greater competence than their male colleagues and superiors — naturally, given their greater familiarity with the day-to-day operation of these devices.

In the days of the typewriter, the photocopier and the filing cabinet (highly gendered technologies from which men kept their distance), this was not a problem. Men had no desire to become familiar with these implements, for there were specialized, subordinate office workers for this purpose. Indeed, men fostered their own ignorance of these technologies in order to successfully maintain this distance, eschewing, for example, the operation of keyboards lest they be seen to be performing a 'low-grade' function. With the increasing diffusion of computerized information technologies into workplaces of all kinds, this ignorance has become an increasing problem for male office workers.

Word Processing in the 1990s: Shifts in the Gendered Division of Labour in the Office

With the spread of microcomputers and business packages like spreadsheet systems Visicalc and Lotus 123, the all-purpose office computer has become increasingly popular. This development has partly arisen from the social relations of power and expertise around the computing specialism. Managerial end users in organizations have challenged the control of centralized DP departments in organizations, demanding greater independence from erstwhile IT professionals and pressing for greater autonomy over their own local and individual information handling activities. The growing popularity of microcomputers and their applications can be seen partly as a function of this struggle over the domain of computing activities and expertise. An increasing range of business applications has been made available for 'stand-alone' microcomputers, including word processing packages like Word-Star (which, as we have seen, initially emerged in a quite different context and for a quite different application but which have markedly improved in sophistication). As this convergence of information technology applications into a single device has occurred, so the popularity of the specialized word processing machine which is dedicated to one function only, has declined. Since the mid-1980s, dedicated word processing machines as separate pieces of hardware have been growing increasingly rare and are forecast to disappear altogether (Vinnicombe, 1980; Hart, 1984). The trajectory of development of word processing has changed direction.

With the rise of the all-purpose office computer, the specific and exclusive association of keyboard operating with degraded women's work has become weaker. As word processing has become associated less with dedicated electronic replicas of typewriters and more with general computing, it has become more acceptable for men to use these devices with dexterity. No longer is keyboard operating the sole province of women office workers. On the contrary, courses in keyboard skills are now specifically designed to attract male managers. For example, Sight and Sound, the organization which runs typing courses, has launched 'Breakfast Time Training for the Busy Executive'; a four week course in keyboard skills aimed at 'putting the businessman in control of his [*sic*] computer' (*The Scotsman*, 20 February 1990). Even secretarial studies courses have been renamed 'Office Studies', so that men as well as women many acquire these and other office skills without the association with 'women's work' (and therefore presumably without embarrassment). Word processing, the activity and the technology, is shifting in its domain, and the boundaries of the old gendered division of labour in the office are being redrawn.

Is this loss of a monopoly of expertise significant for women? Skilled male workers guard such monopolies jealously, for to lose them means a significant loss of occupational power and status (Penn, 1982; Cockburn, 1985). This is not the case for women office workers. Their traditional monopoly of expertise and skill in handling office technologies has not afforded them equivalent levels of power and status at work. On the contrary, as we have seen, office work and women office workers have been ghettoized and down-graded. Indeed, the very concept of 'skill' has sat uneasily with women's work, precisely because of the cultural devaluation of the latter.

As word processing loses its exclusive association with this highly gendered and consequently degraded set of competencies, where does this leave the women who traditionally performed this work? In the short term, it is clear that little has changed in the way that the majority of offices are run. Despite the fact that some men can now touch-type, and no longer feel compelled to maintain their ignorance of how to operate office technologies like word processors, the kinds of activities to which they are applying their new-found skills are on the whole very different from the ones carried out by the traditional secretary or typist. While they may now use their machines to enter text or to operate other types of business software, such as spreadsheets, they tend to perform these tasks in support of their primary activities rather than as wholesale, intensive functions in their own right. That is to say, professional men are taking on only some elements of the text production task. There are still very few men whose entire jobs are concerned with processing other people's words, as do secretaries or typists. Also, as Huggett (1988) has complained, this shift in the division of labour has not released women office workers to the more creative and responsible areas of office work. It has invariably simply left them performing a set of more piecemeal and disjointed word processing activities than before, for example, entering corrections to drafts or tidying up the layout of documents. This indicates that more than technological change is required to change patterns of working and the traditional positions of workers. Organizational and political changes in the workplace, whether of a Taylorist or of a progressive nature, are also necessary; these are much more complex, conscious and long-term processes, as we shall see.

For the moment, the traditional boss/secretary relationship and the associated patterns of knowledge and expertise remain intact in many workplaces; evidence collected by Silverstone and Towler (1983) of the historically changing composition of secretarial tasks bears this point out. Even in typing pools, where the women require less organizational understanding, they still use some technical competence and skill in the course of their work. That is to say, they are continuing to exercise real competencies and abilities at work, despite pessimistic predictions of the deskilling effects of word processing made during the 1980s, and despite the difficulties inherent in attributing to these competencies 'skill' in the conventional sense. An item on the BBC Radio 4 'Today' programme on 5 March 1990 illustrated the continuation of women's skills in the office and their continued importance for maintaining the smooth running of an organization. The item concerned Jim Hodgkinson, a manager of a branch of B&Q, a DIY chain in the UK, and Julie Andrews, his PA, who swapped jobs (though not salaries or promotion prospects) for a day. Whereas Julie found no difficulty in taking over Jim's job, simply executing the tasks which previously she left in preparation for him to execute, he was totally unable to cope with hers. The technologies defeated him in particular. 'The technology beat me, I'm afraid', he said, 'The word processors and facsimile machines and technology dotted around the place were just mind-boggling.'

In the long term, it seems that we need to develop new arenas into which women office workers can move. As Huggett has suggested, the spread of microcomputers in the office provides secretarial workers with opportunities to claim responsibility for a range of activities (for example, desk-top publishing or database management) which would broaden their jobs and expertise. And there have been initiatives to broaden the remit of clerical workers more generally, and to allow them to develop a range of computer-related skills such as systems support or administration. However, this requires a recognition of women's competencies and abilities, rather than a denigration of them. Some analyses have, unfortunately, contributed to this denigration and devaluing of women's office work by highlighting its unmitigated subservient routine and the lack of any real technical competence necessary to carry it out. As such, their over-pessimism has been unhelpful for an understanding of the dynamics and relations at work in the automated office. In fact, the interaction of gender, role and hierarchy seem to govern the pattern of office processes and social relations of control, and these forces complicate the effects of technological change. The application of technology to women's work has quite different connotations from its application to men's work; managements do not usually set about the deskilling of these jobs with the same urgency as they might apply to men's jobs because women's work is seen as already being unskilled. In fact, as we have seen, women persist in their exercise of competencies in the course of their work. This, however, is mediated by their role in the office, with routine workers being confined to the exercise of primarily technical abilities, but those higher up the secretarial hierarchy bringing a wider range of organizational and other competencies to bear.

The interplay of gender and labour process also shapes the direction of technological innovation such that there is a complex and mutual interaction process at work. The entire history of word processing technology can be seen as resulting from the distribution of typing skills and gendered office tasks. The need to reformulate these activities and overcome women's monopoly of expertise has prompted a shift from dedicated machines that mimicked typewriters to an all-purpose technology based on the conventional computer. This suggests that the relationship between technologies, patterns of work organization, and social relations in the workplace is not a simple one, but involves mutual shaping and reshaping. Technologies emerge in particular social contexts in response to particular problems, influencing these and themselves, and are reshaped in response to the changing patterns of work organization and social relations within which they are situated.

Acknowledgment

I am most grateful to Mary Jennings; her insights and expertise contributed greatly to the development of some of the ideas in this chapter.

Notes

1. Henriques and Hoskins (1984) and the Women's Voice Word Processor Pamphlet (1979) are exemplars of 'optimistic' and 'pessimistic' versions of this approach.
2. For examples of these very generalized statements about the impact of IT (particularly word processing devices) upon office labour processes and patterns of work organization, see especially E. Bird (1980), J. Barker and H. Downing (1980), S. Curran and H. Mitchell (1982), and U. Huws (1982).
3. Studies which have shown the complexity of information technology implementation in the office include Arnold *et al.* (1982), Tijdens *et al.* (Eds) (1989) and Webster (1990b).
4. Fleck advances the concept of 'innofusion' to suggest that the implementation of systems represents a crucial site for innovation, and that this feeds back into the development of future systems; see Fleck (1988).
5. This suggests that technology cannot simply be seen as hardware, for it also encompasses social, economic and organizational elements which shape it. See Hill (1981) and Webster (1990a) for examples of this reconceptualization of 'technology'.
6. This later became the Office Products Division. For a detailed account of this chapter in the history of technical change, see May (1981).
7. A number of people advanced this argument, for example, Barker and Downing (1980), Huws (1982), and Women's Voice Word Processor Pamphlet (1979).
8. See particularly the work of Hales (1988), and research reported in Tijdens *et al.* (1989).

References

Arnold, E. *et al.* (1982) *Microelectronics and Women's Employment in Britain*, Science Policy Research Unit Occasional Paper No. 17, University of Sussex.

Barker, J. and Downing, H. (1980) *Office Automation: An Essential Management Strategy*, London, Heinemann.

Benet, M.K. (1972) *Secretary: An Enquiry into the Female Ghetto*, London, Sidgwick and Jackson.

BIRD, E. (1980) *Information Technology in the Office: The Impact on Women's Jobs*, Manchester, Equal Opportunities Commission.

COCKBURN, C. (1985) *Machinery of Dominance: Men, Women and Technical Know-How*, London, Verso.

CURRAN, S. and MITCHELL, H. (1982) *Office Automation: An Essential Management Strategy*, London, Heinemann.

DAVID, P.A. (1985) 'Clio and the Economics of QWERTY', *Economic History*, **25**, 2.

DELGADO, A. (1979) *The Enormous File: A Social History of the Office*, London, John Murray.

FLECK, J. (1988) *Innofusion or Diffusation? The Nature of Technological Developments in Robotics*, Edinburgh, PICT Working Paper, No. 4.

HALES, M. (1988) *Women: The Key to Information Technology*, London, Briefing Pack for London Strategic Policy Unit.

HART, M. (1984) 'How the Office of the Future is Shaping Up', *CA Magazine*, August.

HENRIQUES, N. and HOSKINS, T. (1984) *How to Survive the Office of the Future*, London, Quiller Press.

HILL, S. (1981) *Competition and Control at Work*, London, Heinemann.

HUGGETT, C. (1988) *Participation in Practice: A Case Study of the Introduction of New Technology*, Watford, Engineering Industry Training Board.

HUWS, U. (1982) *Your Job in the Eighties: A Woman's Guide to New Technology*, London, Pluto Press.

MAY, F.T. (1981) 'IBM Word Processing Developments', *IBM Journal of Research Development*, **25**, 5, September.

PENN, R. (1982) 'Skilled Manual Workers in the Labour Process, 1956–1964', in WOOD, S. (Ed.) *The Degradation of Work? Skill, Deskilling and the Labour Process*, London, Heinemann.

PHILLIPS, A. and TAYLOR, B. (1980) 'Sex and Skill: Notes Towards Feminist Economics', *Feminist Review*, 6.

PRINGLE, R. (1988) *Secretaries Talk*, London, Verso.

SILVERSTONE, R. and TOWLER, R. (1983) 'Progression and Tradition in the Job of Secretary', *Personnel Management*, May.

STACEY, M. (1960) *Tradition and Change*, Oxford, Oxford University Press.

THOMPSON, P. (1983) *The Nature of Work*, London, Macmillan.

TIJDENS, K. *et al.* (Eds) (1989) *Women, Work and Computerisation: Forming New Alliances*, Amsterdam, North Holland.

VINNICOMBE, S. (1980) *Secretaries, Management and Organisations*, London, Heinemann.

WAJCMAN, J. (1991) *Feminism Confronts Technology*, University Park, PA, Pennsylvania State University Press.

WEBSTER, J. (1990a) *The Shaping of Software Systems in Manufacturing: Issues in the Generation and Implementation of Network Technologies in British Industries*, Edinburgh, PICT Working Paper, No. 17.

WEBSTER, J. (1990b) *Office Automation: The Labour Process and Women's Work in Britain*, Hemel Hemstead, Harvester.

WOMEN'S VOICE WORD PROCESSOR PAMPHLET (1979) *Job Massacre in the Office*.

Section IV

New Initiatives in Europe and Scandinavia

Chapter 8

'City Libraries': Human-Centred Opportunities for Women?

Eileen Green, Jenny Owen and Den Pain

Introduction

In our introduction to this book, we argued for the importance of integrating gender perspectives within the expanding area of interdisciplinary research on office information systems analysis and design. Then in the first chapter we discussed the area of 'Human-Centred Systems' (HCS) research, in the context of broad trends within computer science, taking the defining features of Human-Centred Systems projects to be:

- — an emphasis on human labour, knowledge and skills, both at the individual and the collective level. Information systems should be designed to complement and enhance these, not replace them; not only in order to protect working conditions, but also in order to produce robust and reliable systems;
- — a recognition that achieving this emphasis in practice requires challenging existing inequalities of power — whether between technologists and non-technologists (Rosenbrock, 1989), or between management and labour (Cooley, 1987).

To date, HCS research in general has not focused on gender divisions as a dimension of power, within or outside of the workplace.[1] The implicit adoption of skill definitions related to male-dominated craft traditions has tended to prevent HCS research from focusing on areas of work in which many women are employed, such as office work. On the other hand, HCS has addressed issues which have much in common with the concerns of those researching gender relations within science and technology. These concerns include a rejection of the opposition between thought and emotion (or 'objective' and 'subjective' knowledge); instead, these are seen as interdependent (Gill, 1990; Rose, 1983). Another common concern is the view that technological development does not proceed in a linear manner; in general there is no 'one best way' to resolve a given problem, but instead a range of contending solutions (Rosenbrock, 1989; Gill, 1990; Haraway, 1991; Bijker, Hughes and Pinch, 1987). There is therefore considerable potential for constructive exchange between HCS research and research on gender and information technology.

In this chapter, we reflect upon our own efforts to integrate gender perspectives within an 'action research' project on human-centred office systems design. This was conducted between 1987 and 1991, in close collaboration with management, staff and trade union representatives in a major UK public library: 'City Libraries'.[2] Both the organization and the research process are described in detail in the second section of this chapter. In the later sections we discuss in depth issues concerning gender and user-involvement. At the outset, however, it is worth highlighting a number of tensions in each, which proved both creative and constraining.

First, during this period the research team itself included two computer scientists, both men, and two social scientists, both women. We were all committed to attempting to share skills and to develop new, interdisciplinary perspectives; however, each of us also brought our own uncertainties and ambivalences to the project, particularly since our own membership reflected such a familiar, gendered pattern. It seemed very clear from both published research and our own experience, that many computer science methods for the development of computer systems had impacts on organizational structure, the quality of work and social groupings at work. So if human-centred developments were to succeed, then expertise from social science was required to balance the inadequacies of computer science in organizational contexts.

On the computer science side, there was a concern not to allow 'technical' know-how, or computer hardware and software, to dominate. From a social science perspective, however, this genuine concern sometimes appeared to devalue precisely the skills and the knowledge to which many women (including the two on the research team) were feeling an acute need to open up access on better terms.

In a more general sense, the project (like other HCS research) had originally been influenced by Braverman's deskilling thesis (1974), and our decision to focus on women office workers, rather than on men in manufacturing, soon began to raise gender issues which could not easily be accommodated within a labour process perspective (Davies and Rosser, 1985; Beechey, 1987; West, 1990). In a sense, then, the research project itself came to embody a number of stages within a more general shift which was taking place within feminist theory: a shift away from 'adding women on' to existing theories or research objectives, and towards theorizing gender relations within and beyond the workplace, in terms of both women's and men's experience (see for instance Cockburn, 1985; Harding, 1986).

Second, both City Libraries and 'Northfield Council', the local authority of which it was a department, offered constraints as well as opportunities for the research project. There were many very positive factors. This council was one of a number of similar organizations which developed radical policies during the 1980s. Generally speaking, these policies were intended to promote local democracy, and the development of innovative public service provision. Inevitably, they brought these Labour-controlled local authorities into direct conflict with a Conservative central government whose own policies were constructed upon the notion of individual choice and effort, the privatization of publicly owned resources such as energy and transport, and 'market values' in general.

Northfield Council, for example, had initiated schemes to promote local training and employment, and had established specialist units to address inequalities of race and gender. The council had also supported both a Women's

Technology Training Workshop and a community-based initiative to create access to information technology use and training for unemployed people. These initiatives shared a number of the concerns which have characterized HCS, notably for instance the focus on shaping technology to fit employment or community needs, rather than vice versa (Cooley, 1987). Such a context could, thus, offer some potential for new approaches to information systems development.

The Case-Study Context: Positive and Negative Features

Interestingly, however, these innovative approaches to IT were not reflected in Northfield Council's own internal strategies and practices. IT strategy was largely limited to a commitment to buying equipment from one main supplier (ICL). New 'bids' for IT development were discussed and approved by a panel of councillors who generally relied upon the advice of senior computing personnel. The terms of debate were 'technical' and financial in the narrowest senses, marginalizing any consideration of organizational issues and implications. In this context, there appeared to be no links or dialogues being established between policies on IT and those on issues of gender (or other) inequalities. The innovative training and childcare models developed by the Women's Technology Training Workshop, for example, had not been disseminated through other forms of local authority provision.

City Libraries itself was a small department in the council context, but as a major public library it had taken a leading role in policy and service development, both locally and in the national library context. This included playing a substantial part in the establishment of the 'Libraries Campaign' in 1987; the aim of this was to increase public support for libraries at a time when many were experiencing major cuts in resources. In addition, following the appointment of a confident and creative woman director in 1983, City Libraries began to develop a model of librarianship which emphasized community links and outreach work: an expansion of the traditional book-lending and reference role. This in turn required a higher degree of staff development and consultation. As part of this evolution, from 1984, management took up a trade union proposal to improve pay and career development opportunities for non-professional library staff (library assistants) on clerical grades. Broadly speaking, through this 'alternative grading' scheme, management recognized that there was a degree of overlap between library assistants' skills and experience, and those of their professional colleagues. At this point, 395 of the 400 library assistants at City Libraries were women, while men predominated in the senior professional and managerial posts. The alternative grading initiative therefore represented an important move in the direction of improving women's pay and status within the organization.

As within the broader local authority, however, these new initiatives were not reflected in City Libraries' approach to IT development and use. By the mid-1980s, computerization was a major source of concern, both for management and for professional and non-professional staff. At this point, three 'generations' of library information systems coexisted at City Libraries. A small number of branches and sections still operated entirely on a manual basis, using card-catalogues and issuing books in return for tickets under the traditional 'Browne' system. The majority of the service relied on a batch-processing computer system, housed on the City Council's mainframe computer. This produced a microfiche catalogue, updated every three months, and processed book issue and discharge and borrower details. This system had been constructed piecemeal over a ten-year period, and in fact had never been completed satisfactorily. Its many shortcomings and peculiarities were the source of great frustration for staff, one of whom summed it up as 'an iron carthorse with ribbons'. Finally, in a small number of sections, staff had on-line access to specialist databases outside City Libraries; in some instances, staff had also constructed their own PC-based databases for particular purposes (for example, to catalogue the drama collection and other specialized resources).

Experience of these newer facilities, combined with rising public expectations, led management in particular to see the acquisition of an on-line, integrated library system as the way to sustain service provision, especially in a context of static or shrinking funding for local services. A system of this kind could be bought, in its basic form, from a specialist supplier; being largely parameter-driven, it could then be adapted to local requirements. Such a system was not expected to facilitate job loss, but to enable both service provision and working conditions to improve.

A key concern on management's part was to undertake 'consultation' with library staff in relation to new phases of computerization, since they attributed many of the failings of the batch-processing system to the total lack of user-involvement during its planning and design. Both professional and non-professional staff in City Libraries shared this opinion. In addition, the partial erosion of the historically sharp divide between professional and non-professional library work suggested a good basis for clerical involvement in systems development. However, neither management nor staff had a clear model of how they might promote user-involvement in the acquisition of a new system. In this context, when we approached City Libraries with a view to initiating collaborative research on new approaches to user-involvement, interest was expressed both by staff and by management. However, there was also considerable caution, paraphrased in the following comments:

> We'd love to get involved, but how do we know that management will listen to what we say?

> You're welcome to try this, but you'll find the staff very apathetic . . .

To sum up then, three aspects of City Libraries as an organization created a good basis for the exploration of human-centred systems approaches. First, among both management and staff there was a commitment to improving service delivery and working conditions; this included an increasing recognition of gender issues (in the context of wider concerns with equality in employment and in service provision). Second, the scope of clerical posts was being reviewed, and some new opportunities for career development were being opened up. Third, existing information systems (both manual and computerized) were seen as increasingly inadequate, and both staff and management saw 'user-involvement' as essential to new phases of computerization.

In the following section, we describe the main features of City Libraries as an organization, and the collaborative research undertaken with staff and management there. In the third section of the chapter, we then turn to a discussion of the main themes which emerged in the course of this research, in connection with gender relations and human-centred systems approaches.

City Libraries: The Starting-Point

With a workforce of 32,000 when our case-study began in 1987, Northfield Council was by far the largest local employer. City Libraries itself had some 520 staff; the service was composed of a network of thirty-six branch libraries and a Central Library housing both the most heavily used public lending library in Britain and a number of specialist reference sections. Reflecting the pattern within local authority employment as a whole, most manual, clerical and administrative staff in City Libraries itself were women, whereas men predominated at senior professional and managerial levels (Stone, 1988). This has not always been the pattern within library employment, however; women out-numbered men, as professionals, before graduate entry became the norm:

> It used to be that most of the senior staff were women, who moved up through local libraries. Now we've got more people who have gone through the professional channels, and a higher proportion of men; that has worked against women, in the last ten years, particularly at middle management level. (Neil Jenkins, Bibliographical Services Deputy Section Head and chief shop steward)

During the 1970s and early 1980s, there had been little incentive for women library assistants to take up further training. For many, gaining a full professional qualification would have required giving up work to study full time; part-time City and Guilds courses were available, but taking these offered no reward within a grading structure in which only one basic grade was available. With the introduction of the 'alternative grading scheme', however, more library assistants had already begun to take up a range of courses on day-release.

We have already indicated the unanimous dissatisfaction with existing computerization that prevailed at City Libraries in the mid-1980s. Early interviews with library assistants and cataloguing staff graphically illustrated how inaccessible the systems development process had been to any of the eventual users of the batch-processing system. They also emphasized how stressful and demanding this system could be in practice:

> At a fairly late stage, we started having regular meetings with Computer Services staff, and a series of trial runs was established over a year, based on sample databases. But we couldn't do genuine random testing until it was too late to turn back. With most problems that emerged, we were told that nothing could be done about it. (Neil Jenkins)

One library assistant described the 'obstacle course' represented by the resulting layers of catalogue facilities in the following terms:

> There are four parts to the catalogue system now: the card catalogue up to 1970; the card catalogue from 1970 to 1985; then the microfiche, and

> then the on-line short title file . . . You have to be more careful, more knowledgeable, than with the card catalogues. It's not 'user friendly'; there's no help screen, and often you're just left in the lurch to muddle through. (Josie Walker, reference library assistant)

Recognition of these difficulties, and of the fact that the existing systems could not accommodate the increasing volume and sophistication of library transactions, led the departmental management team to envisage the eventual acquisition of an on-line, integrated library system. In the mid-1980s, some eight to ten suppliers were beginning to market systems of this kind, which are constructed around one large relational database. This can support the whole range of library activities: from ordering and cataloguing materials, to issuing and discharging them; dealing with all borrower and user enquiries; and generating management reports and budget information.

In 1985, two management team members had been delegated to prepare a draft systems specification, working in collaboration with an 'IT Policy Development Group' composed of managers and senior professional librarians. But by late 1986 it had become clear that this work was being continually deferred; policy development group members were all over-committed in other directions. Senior management therefore welcomed our research approach. Following discussions with both the management team and the trade union branch, we were able to reach an agreement to collaborate, based on three principles: first, a view of library automation as enhancing jobs and services, not replacing staff; second, a view of the systems planning and development processes as embracing both technical and organizational aspects; and third, a commitment to exploring new forms of consultation or involvement for women library assistants on clerical grades. Overall responsibility for managing the process remained with the libraries' departmental management team.

Two New Approaches: Study Circles and a Broad-Based Design Team

Our research began with a series of preliminary interviews with staff, trade union representatives and managers in the Central Library. These revealed the degree of ambivalence and uncertainty (concerning future phases of computerization), to which we have already referred. We therefore needed a form of initial approach which would enable female library assistants to articulate their own observations and questions regarding IT. Many were already intensive systems users, but none had had access to the departmental discussions concerning possible new systems. We were acutely aware that few women in City Libraries (professional or non-professional) regarded themselves as having acquired IT expertise; at least, not in the same sense that it seemed to have been acquired by small groups of men at more senior levels, who could be seen 'having a play' with new bits of software or PCs at their desks. We return to this issue in the final section.

In this context, the 'study circle' method appeared to us to be the most appropriate starting point. Workplace study circles have been used as part of the Scandinavian 'Collective Resource' approach, briefly discussed in the first chapter. Swedish trade unions have used them as part of consultation and training processes in connection with the introduction of information systems, and they have also been used as part of British trade union initiatives (unconnected with IT) to make contact with women members in the service sector (Olerup *et al.*, 1985; Avner, this volume; Vehvilainen, 1986 and 1991).

In contrast with management-inspired 'quality circles', study circles emphasize active involvement by the participants in working on an agenda which they define themselves. The groups can provide a basis for the process which feminists would call 'consciousness-raising': sharing and comparing experiences in order to raise questions and develop new understandings. They can lay a basis for networks of mutual support, and enable participants to gain access to new ideas, contacts and information.

We viewed the study circle method as appropriate in the City Libraries context because it offered scope for library assistants to take part in our research, and to innovate in ways we had not foreseen (Silverman,1985). The study circle process invites participants to identify issues and work on them, through practical activities and through critical reflection. It is a means for arriving at new ideas and plans, rather than for putting across a pre-existing model or body of knowledge.

In terms of HCS, then, we wanted to assess the study circle technique in two respects. First, we were concerned to discover whether, as a systems development technique, it might offer a good basis for genuine partnership and collaboration to develop between 'users' and 'experts', particularly in the early stages when priorities and requirements are being defined (Ehn 1988). Second, from a gender perspective, we wanted to invite women library assistants to review their own working knowledge and experience, and to use it actively in assessing new library system possibilities. Previous research on gender and technology has indicated the ways in which 'technical expertise', often in the hands of men (although not inevitably), tends to displace women's knowledge from having 'authoritative' status (Suchman and Jordan, 1989).

The creation of a broad-based design team to oversee the acquisition of a new computer system emerged directly from the study circle meetings. Members of the first study circle proposed that this team should be established in order to carry out detailed work on the planning, specification and implementation of a new system. This represents an early example of female library assistants' assertive intervention. They were aware of the need for a coordinating group which would be both more flexible and more representative than the former 'IT policy development group'. Accordingly, the new design team initially included three library assistants working alongside colleagues from middle management, computing and other professional posts. Motivation among study circle members to be part of this group was high, since they perceived it as a channel through which their own priorities for new phases of computerization could be made explicit for the first time. Observation of the team's work offered a unique opportunity to assess the scope for women clerical workers' involvement on a full basis, rather than on the much narrower basis associated with the more conventional systems development methodologies discussed in the first chapter.

A Sequence of Activities

The first study circle met in the spring of 1987. In the following September, study circle members presented a formal report to the departmental management team.

For the first time, this brought senior management and clerical workers face-to-face to discuss what should happen next. The management team agreed to the two substantive proposals put forward in the study circle report: first, time and resources to enable more library assistants to take part in study circle groups, and second, the establishment of the broad-based design team briefly described above. During 1988, four more study circle groups met; these were followed by two more in 1989–90, so that eventually over 10 per cent of library assistants in the organization had taken part.

Clearly, then, the study circle process gained considerable momentum during 1988 and 1989. In contrast, as a body with the task of tackling unfamiliar ground in new ways, the design team found itself struggling to define a way forward during its first year. Communications with the senior management team had never been consolidated; both sides were unclear about their roles and responsibilities. In addition, the three library assistant members began to feel increasingly isolated; their contributions were becoming dissipated, largely because of the prevailing uncertainties in the group. This situation exposed them to criticism from their immediate colleagues, suggesting for instance, that their participation was tokenistic and ineffective.

In response, our research team organized a seminar early in 1989, designed to bring together all those concerned with the project: researchers, City Libraries senior management, study circle members and design team members. Members of the research team played a facilitating role, providing a non-library forum in which to analyze the reasons for lack of progress and to air the building resentments before they could become destructive. This meeting helped to empower the library assistants by putting them in a majority and ensuring that during policy-making workshop sessions they were in a group on their own.

Outcomes of the seminar included defining the work as a 'Computers and People Project', which was composed of an expanded design team and a new 'support group'. The former now included seven library assistants, nearly half its membership. At the request of library assistants, it was to be chaired by Mo Smart, a senior management team member, in order to ensure effective communication with the management team.

The support group was intended to draw together a number of people whose roles would enable them to maintain an overview of the detailed work of the design team. It included key staff from the areas of training, research and personnel, together with a trade union representative, the computer development officer, members from our research team and the chairperson. At the same time, a number of small sub-groups were convened to carry out detailed work in areas such as job design, health and safety and training. Coordinated by design team members, these also drew in other professional and non-professional libraries staff (see Figure 8.1).

This research team intervention was an essential element in a general revitalization of the process; it created a space within which design team members could find ways to sharpen both the nature and role of the group, and in particular, improve communications with management. This does underline the importance of a 'facilitation' role in relation to processes of organizational change and development. However this role may be interpreted in many ways; Socio-Technical Systems approaches, for instance, suggest a fairly conventional consultancy model (Mumford, 1987; Hirschheim, 1985). In contrast, the Scandinavian Collective

Figure 8.1: Post-seminar organization

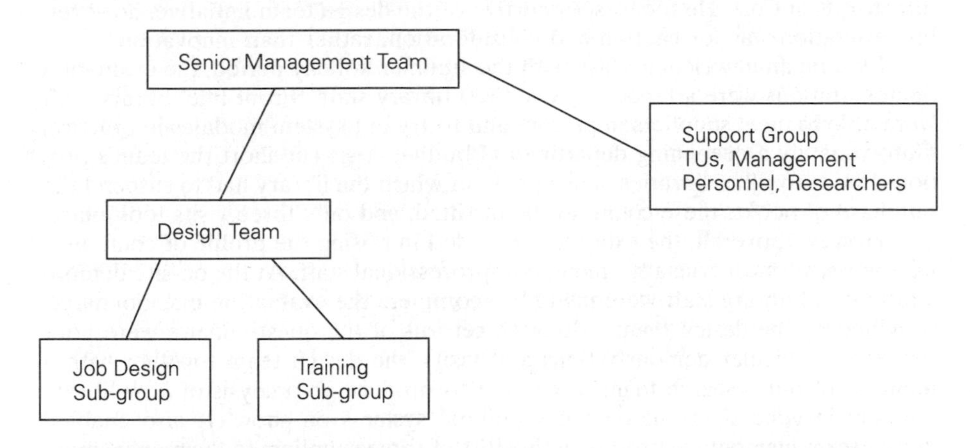

Resource approach argues for a partnership in which systems designers, users and social scientists aim to share knowledge and skills (Ehn, 1988). This second emphasis was the one for which we were aiming.

On this basis, progress was re-established. The design team held a seminar for prospective systems suppliers in the summer of 1989. This meeting was organized very much on the design team's terms; that is, if computer systems suppliers were to attend, they had to agree to both initial and ongoing involvement by the library assistants. This agreement highlighted the organization's commitment to a human-centred development approach, based on involving 'users' at all levels. This event enabled the design team to grasp the initiative in relation to suppliers and to formulate initial impressions about their products and their attitudes to user-involvement. The team then began to make detailed plans for evaluating both suppliers and systems, through visits to other libraries and through 'hands-on' demonstrations on City Libraries' premises. The draft systems specification was completed for circulation and discussion in December 1989. During autumn and winter 1989–90, detailed evaluation questionnaires were drafted for use in systems visits and demonstrations. Library assistants drafted the initial questions for these documents, which were then expanded and amended by the whole team.

By now, design team members had gained in confidence and experience, and were looking forward to the period of practical evaluation activities. However, both the team's chairperson and the computer development librarian responded with considerable ambivalence to this increased assertiveness — an assertiveness which represented a new readiness, on the part of professional staff and of library assistants, to share in the week-to-week management of the project. They gave team members very little information about the process of liaising with systems suppliers, and did not take up their repeated offers to share in this work. The resulting delay meant that evaluation visits and demonstrations finally took place in the summer of 1990, some four to five months later than the team had originally planned. This difficult period coincided with a change of departmental director; Penny Cook, whose innovative policies had underpinned the study circle initiative, had left in the summer of 1989, to be replaced by the former deputy director, Ken Cole. He too was supportive of the design team initiative, however, his reputation was for caution and consideration, rather than innovation.

Despite an unwelcome clash with the summer holiday period, the evaluation demonstrations were attended by over 100 library staff. 'Front-line' library staff were able to meet suppliers in person, and to try out system modules in practice. Unfortunately a deepening departmental budget crisis cut short the team's proposed visits to other libraries; in a climate in which the library had to suspend the purchase of books, these could not be justified, and only three visits took place.

However, overall, the exercise succeeded in raising the profile of computerization issues, most crucially among non-professional staff. At the on-site demonstrations, all library staff were invited to complete the evaluation questionnaires provided by the design team. Although returns of the questionnaire were poor for some particular demonstrations and visits, the design team together with a member of our research team, were able to produce an analysis of how library workers in general evaluated the suppliers' systems on show. It also enabled design team members to propose a shortlist of three suppliers to the management team in December 1990. This, combined with the circulation of the draft specification, stimulated increased interest from union shop stewards in the library, who formed their own group to monitor progress.

By early 1991, the general budget crisis within the local authority had forced a programme of voluntary redundancies across all departments. City Libraries had to lose forty staff, some 8 per cent of the total. Among those who chose to leave was Mo Smart, the design team chairperson. Demoralization was widespread. Although the departmental director received assurances that a capital budget was still set aside for the purchase of the new system, lack of resources for relief staff began to undermine library assistant participation in the design team. Mo Smart was not formally replaced until the summer of 1991, by a senior male manager, Clive Monks.

Clive Monks took the opportunity to lead a review of the process, resulting in considerable clarification of roles and activities (see Figure 8.2). In terms of library assistant participation, this outcome had positive and negative aspects, which we discuss further below.

As originally planned, the research team withdrew from active involvement at this stage, and convened a workshop at which an assessment of the process was discussed with both the design team and senior management team members.

Discussion

In reflecting upon the period of action research summarized above, three themes are of particular interest: first, issues of gender in relation to user-involvement; second, the strengths and weaknesses of the two design initiatives adopted; and third, two sets of unresolved tensions exposed by the research.

Gender and User-Involvement in Office Systems Design

There is now considerable research evidence to support Friedman and Cornford's argument (1989) that 'user relations' concerns are increasingly prominent in

Figure 8.2: Post-review organization

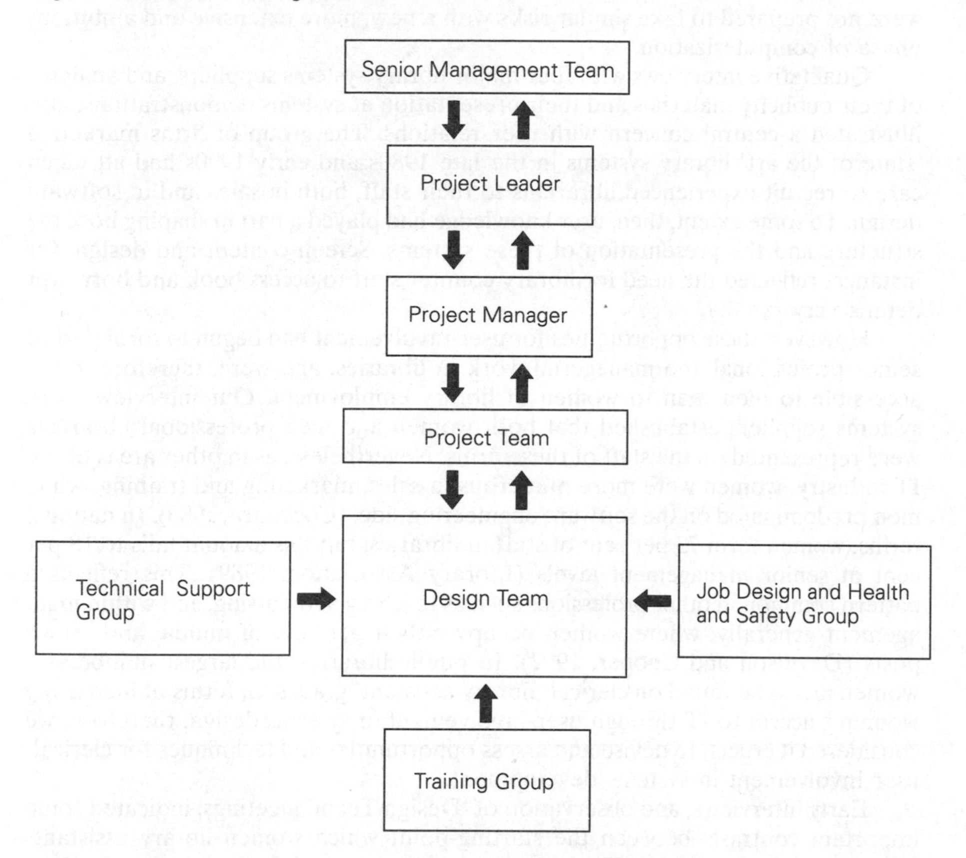

connection with information systems development. In the office context, our own earlier case-studies illustrated this, but also suggested that gendered hierarchies in organizations were reflected in these trends. In a major engineering firm in the private sector, we found that a prototyping approach had been successfully adopted; clerical users' involvement contributed significantly to the success of this project, but remained informal and largely unacknowledged. Also in a large local authority personnel department we observed the involvement of professional users through the adoption of a version of the Structured Systems Analysis and Design Methodology (SSADM). These users retained much of the initiative, but issues of organizational development and change, most notably clerical job design and VDU use, were marginalized by a narrow focus on design work as a 'technical' activity. This focus undermined the viability of the project as a whole (for a detailed account, see Owen, 1992; and Green, Owen and Pain in Woolgar and Murray, forthcoming).

Our research at City Libraries allowed us to examine this general trend towards increased user involvement in greater depth. City Libraries' own management team was prepared to consider involving users in the planning and acquisition of a new system, because they explicitly attributed past systems failures and inadequacies to the previous exclusion of users from these processes. They were not prepared to take similar risks with a new, more extensive and ambitious phase of computerization.

Qualitative interviews with four major library systems suppliers, and analyses of their publicity materials and their presentation at systems demonstrations, also illustrated a central concern with user relations. The group of firms marketing 'state of the art' library systems in the late 1980s and early 1990s had all taken care to recruit experienced librarians to their staff, both in sales and in software design. To some extent, then, user knowledge had played a part in shaping both the structure and the presentation of these systems. Screen content and design, for instance, reflected the need for library counter staff to access book and borrower details very rapidly.

However, these opportunities for user-involvement had begun to form part of senior professional and managerial work in libraries, and were, therefore, more accessible to men than to women in library employment. Our interviews with systems suppliers established that both women and men professional librarians were represented on the staff of these firms. Nevertheless, as in other areas of the IT industry, women were more numerous in sales, marketing and training, while men predominated on the software engineering side, (Cockburn, 1985). In national terms, women form 73 per cent of staff in libraries, but this amount falls to 18 per cent at senior management levels (Library Association, 1989). This reflects a pattern common in other professions such as teaching and nursing, and within management generally, where women occupy only 4 per cent of middle and senior posts (Davidson and Cooper, 1992). In public libraries, the largest numbers of women are to be found on clerical 'library assistant' grades. In terms of increasing women's access to IT through user-involvement in systems design, therefore, we considered it crucial to devise and assess opportunities and techniques for clerical-user involvement in systems development.

Early interviews, and observation of 'Design Team' meetings, indicated some important contrasts between the starting-point which women library assistants might have for discussing IT, and the starting-point of their male professional colleagues and managers. Dinah, a member of the first study circle in 1987, described how she and other new library assistants first encountered 'the computer'. This encounter took place in the mid-1970s, in a manner which clearly constructed the women as 'novices' dealing with technology as something somewhat mysterious and apart:

> I was very overawed, actually, because — when I started — you were sort of kept away from it for the first couple of weeks. And the public — you weren't let loose on the public or the issuing system for the first few weeks. [laughs] You were locked away in the work room to jacket books . . . and we used to practise with — we used to call it a 'spin-drier', it was a tall thing where you put cassette tapes in, and you had to practise on that, you weren't allowed to use the actual terminals to start with. So you thought it was much more complicated than it actually was. Nobody really explained what was happening behind the scenes; you thought that if you did press the wrong button, or you used the pen wrong, then you were going to really mess it up. And it's not the case at all! [laughs]

By the mid-1980s, computer use on the counter had become a matter of routine, and new library assistants were no longer 'hidden away' to practise on the 'spin-drier'. While the connotations of computer use had shifted, however, they had not become any more positive, as Dinah illustrated:

> The counter work has become easier, and quicker. But it [computer use] has taken what little personal contact there was away from the central library. Now it's so quick — you shove people through . . . so I think a lot of people do feel, if they're just starting out, that all they are, are computer operators, they're not librarians or library assistants any more.

In contrast, for the men in the central lending library (professionally qualified librarians) increased computer access appeared to represent a chance to experiment with small specialist databases and with a staff timetabling system.

Dinah continues:

> The men in our section are very sort of enthusiastic about it; anything that comes along new technology-wise, whether it worked or not, was 'super' . . . Suddenly you had to do everything on that piece of equipment. Quite honestly, sometimes we used to think, 'Well, how did we do the job before?' Because they couldn't do anything unless they used that terminal. Having said that, they were both professional staff.

She goes on to describe how women library assistants in her section had refused to operate the PC-based timetabling system set up by their two male professional colleagues, because it was so much more cumbersome to use than the familiar pencil-and-paper chart.

It is important to note here that *women* professional staff were not a passive element in this picture, but the choices they had made had been different. Women professional librarians (including Penny Cook, departmental director) had led all the initiatives on gender inequality (for example, new training opportunities for women) in assertiveness and in management skills. They also facilitated the production of a report on equal opportunities in City Libraries and the negotiation of an improved maternity leave scheme, which included a support network for women on long-term maternity leave. The fact that these areas had appeared to be both accessible *and* urgent priorities for professional women, but not for most professional men, in the department indicates the scalc of change required — a change which needs to take place among both men and women, in order to break down established associations between masculinity and technology, and between femininity and reproduction.

The different expectations and experiences of women library assistants and men in professional posts surfaced particularly clearly in some Design Team discussions when the new systems specification was being drafted. One such example concerned the processing of reminders sent out for overdue library items. This had been a very sensitive issue for library assistants, who were often on the receiving end of complaints when batch-processed reminders went out for items which had already been returned. In the following extract, the Computer Development Librarian (CDL) is responding to various library assistants (LA), who want a distributed system under branch control, rather than a centralized one:

CDL: Why send overdue letters from each service point?
LA: The currency of reminders is important; overdue letters can be a source of friction now.
CDL: But you could put a disclaimer in the letter, in case a book had been returned already.
LA: But for instance, elderly people do get very upset; lots of people get upset.
CDL: But they are a small percentage. You could have a facility to suppress overdues. It's not just a question of automatic printing; it can also collate and stamp, it's all automatic. There are cost implications. Having a large enough stock of paper at each service point would cost a lot of money; and paper might run out in branches.

This extract brings to life some aspects of the gendered dichotomies which have been discussed in a number of studies of information technology and of office work (see for example Vehvilainen, 1991; Bødker and Greenbaum, this volume). Female library assistants' concerns arose in the context of daily routines with 'caring' or 'social' features which have historically been defined as 'women's work'. Their day-to-day experience of systems delays and breakdowns also caused them to feel sceptical about technology. The male computer development librarian was quick to respond with a reference to the 'small percentage' of complaints ('hard' information), and to the image of the streamlined machine as unproblematic, before resorting to the incongrouous suggestion that 'paper might run out'. His responses also echoed familiar gendered patterns (Murray, this volume).

As the above discussion proceeded, the library assistants gradually ceased to take part, exchanging exasperated glances; this was not the first or the only occasion on which they had found their descriptions of their own experiences questioned or rejected, by senior colleagues. However, the final systems specification did embody a compromise: library assistant concern with the level of borrower complaints was recorded very explicitly. This prefaced a requirement that overdues should be processed centrally but with a quick and effective facility for branch staff to monitor and override them. In this and other examples, the involvement of women library assistants did contribute both to articulating their own priorities, and to creating a more comprehensive systems specification than would have resulted without their presence.

Study Circles and the Design Team: Strengths and Weaknesses

Our second concern, in this section, is to examine both the strengths and the weaknesses of the approach we evolved through our analysis of clerical user-involvement at City Libraries. In terms of strengths or achievements, a number of aspects are important. The study circle initiative broke the bureaucratic deadlock which had threatened to stifle effective debate concerning new library systems. It also represented a *de facto* alliance between three 'parties' who all stood to gain from experimenting with new approaches to IT. They included: Penny Cook, the departmental director, who had maintained a deliberate distance from detailed discussions of specific systems and suppliers, in order to retain a clear focus on organizational objectives; women library assistants, to whom previous discussions of IT plans had not been open; and ourselves as researchers.

On the basis of this informal alliance, a number of important changes were achieved. For the first time, large numbers of women library assistants took part in detailed discussions and assessments of IT possibilities, both through the study circles and through the subsequent systems demonstrations. Their views were formally expressed through study circle reports, evaluation questionnaires and contributions to the systems specification. This effectively dispelled the view, formerly held by some senior male managers, that women non-professional staff had become too discouraged by bad experiences of IT to make an active contribution. One library assistant illustrated the ways in which increased management recognition is important, especially on an informal level:

> It [the study circle] certainly made you feel that suddenly other people knew you were around, because you were getting contact with the higher management . . . . just even at coffee breaks, they'd come and sit and talk to you, because suddenly you were a face that they knew. So I mean, you felt much more confident with them anyway.

In addition, new structures were devised to take the systems development process forward, that is, the 'design team', with its network of sub-groups carrying out detailed work. Finally, the study circles contributed to a general acknowledgment that internal departmental communications needed attention. For many participants, the study circle represented the first opportunity to gain a clear view about how their own work related to that in other parts of the library service.

The study circle method also broke another kind of deadlock, since it did enable both participants and researchers to move away from 'taking technology as given' (in the sense discussed by Flis Henwood in Chapter 2 and by Fiorella de Cindio and Carla Simone in Chapter 10). The starting-point was members' own questions, observations and aims, regarding current and future forms of library computerization. Thus their own experience and skills were treated as 'authoritative', rather than as marginal (Suchman and Jordan, 1989). This exposed and challenged the very limiting basis on which most women library assistants had first encountered IT, with all the gendered associations illustrated above. Study circle meetings and subsequent interviews illustrated emerging debates about where 'systems' issues ended and 'policy' issues began. In more theoretical terms, it became apparent that the boundary between 'technical' and 'organizational' domains was not absolute, but was open to negotiation (Williams, 1990; Murray and Woolgar, 1990).

The question of system response times, for example, was a crucial consideration for library assistants; delays of more than a second or two, in penning and issuing or discharging books, could result in unmanageable queues building up at busy times. At first glance, this appeared to be a typically 'technical' matter related to system size and processing power, but study circle visits to other major libraries enabled participants to map out the ways in which organizational priorities had shaped these apparently technical aspects during systems development and implementation. In one major northern public library, book-ordering and cataloguing facilities had to be suspended when the lending section was operating at full capacity; in another, various management information facilities remained unused because the demands they placed on processing capacity degraded response times at the counter so severely. Library assistants on the design team were able to use these observations in order both to raise questions about some of the systems suppliers' more inflated claims, and to negotiate over the ways in which guarantees for response times should be included in the systems specification.

Lastly, the study circle process supports an increasing range of research which suggests that contrary to expectations stimulated by Braverman's labour process analyses (1974), many areas of clerical work have not been reduced to deskilled or routine operations with the introduction of information technology. In some instances, an already fragmented and routinized pattern of work organization has itself shaped and constrained the potential of office information technology (Webster, 1990); in others, clerical jobs are undervalued, but remain relatively skilled and complex (Olerup *et al.*, 1985; Gaskell, 1987). In the public sector in particular, clerical work commonly includes a wide range of activities and responsibilities.

At City Libraries, library assistants' skills included dealing with complex reference enquiries, maintaining specialized indexes and teaching library users to operate new equipment (such as microfiche and public access word-processors). Partial computerization had required an expansion rather than a reduction in their range of skills. Book-issuing became faster and simpler, but searching the catalogue now involved checking and double-checking a range of manual and terminal-based sources. During the study circle process, library assistants were able to put forward detailed proposals for improving approaches to computerization by drawing on their own experience. This was not limited to describing current, day-to-day requirements; both their criticisms of the system they worked with, and their visits to newer forms of library system enabled them to formulate new ideas. For instance, study circle discussions and reports emphasized the complexity of library assistants' enquiry work with library users. In the systems specification, this was reflected in proposals for flexible forms of catalogue access, produced by collaboration between reference library staff and cataloguing staff.

The weaknesses in the City Libraries example emerged most clearly in relation to the activities of the design team. The study circle process created a relatively strong base for initial library assistant participation in the team, both in terms of confidence and of informal networks of support. No parallel process was available to professionals and middle managers to equip them to work alongside library assistants in a new, collaborative way. Our starting-point, in the spirit of labour process theory, had been to concentrate on 'empowering' those with least resources within City Libraries. However, setting this process in motion gradually began to expose the ways in which a 'human-centred' systems development approach would also be dependent on changes in managerial practices.

As the recession deepened in the early 1990s, Northfield Council suffered Conservative Party-inspired, centrally-imposed cuts in local authority spending. This was to have a major impact upon the radical policies developed in the 1980s, not least in City Libraries. A series of 'tightening' of key budgets made it increasingly difficult to implement the changes conceived in a political climate more favourable to local democracy, which led to a deepening cynicism on the part of library assistants and a nervousness among the management team. The change of director in 1989 became linked to the promotion of a more 'market-oriented' philosophy within the organization, which marginalized equal opportunities, albeit unintentionally. This climate formed the background to the efforts of a design team struggling with a complex systems development process.

Tensions within the Design Team

A series of major tensions emerged within the design process, some of a 'slow-burning' nature, and related to a legacy of lack of clarity around equal opportunities issues, and others more immediately connected with the task in hand. An example of the first takes the form of the contradiction between the management commitment to releasing library assistants to take a sustained role in the design process, and the former's tendency to overlook or minimize the effects which the absence of key staff would have on day-to-day work. The assumption that women attending meetings was 'simply' a matter of cover arrangements, and/or squeezing more work into smaller time spaces, represents a naive underestimate of the complexities involved. Library assistant members of the design team, already demoralized by the lack of progress, became further demoralized when faced with a) criticism from peers disrupted by their absence, and b) the increased pressure on their domestic obligations as a result of design team commitments.

The enthusiasm developed as a result of their study circle work was in danger of being dissipated by a lack of energy and clarity from the managers involved in the process. A key example already described above took the form of a reluctance by both the computer development librarian and the chairwoman of the design team to share information about progress and to initiate plans for the next stage effectively.

Management Concerns and Informal Networks

This increasing reluctance on the part of middle managers and some professionals to engage openly with other design team members and accept the offers from library assistants to facilitate action, for example, in arranging systems demonstrations, can be explained on a number of levels. First, there is little doubt that living with the constant threat of externally imposed constraints such as cuts in local government spending, threatened redundancies etc. demoralized a management team attempting to regroup around a new director and a changed market situation for library services. Existing internal problems associated with the transition to a new management culture, born of the early 1990s' economic recession, were deepened by uncertainties about being able to deliver a new system for City Libraries within the agreed time-frame and radical design philosophy. Perceived threats to the ongoing democratic design process resulted in 'entrenchment on safe ground' which took a gendered form. Paul Bates, the computer development librarian, revitalized traditional IT network contacts within Northfield Council, networks characterized by traditional, male-dominated activities such as the provision of informal systems purchasing advice. His situation was complicated by the desire to employ library assistant participation to 'force' changes which he knew to be long overdue, and a conflicting desire to remain secure in his own traditionally, 'technical' role (Cockburn, 1985). His concerns are expressed in the following extracts from an interview with one of the research team:

PB: ... but Central Library is still looking at the way they are operating now, and all they're going to do is swap these new terminals for their old ones and carry on and they can't do that ... the professional from Central Library seems more entrenched that most, and the library assistants are more flexible I think.

... I think the computer system is really a catalyst to something else that's going on, and one of the things that I am worried about from time to time, is [that] this whole idea of library assistant involvement: study circles and so on, seems so closely tied in with this particular project ...

DP: And, therefore, will finish?

PB: Right, it's very much tied to IT and when the system's over, well OK we can carry on because the library assistants can still be consulted in relation to the computer system. But lots of other policy matters and decisions [are] going on in the department, where library assistants have an equal input that's got nothing to do with computing whatsoever.

DP: So ... you see this kind of involvement could be used elsewhere in the library?

PB: Yes, very much so.

The above discussion demonstrates the computer development librarian's commitment to broader organizational changes, which could be achieved on the back of the radical systems design process.

The 'Techies Club' and Forms of Masculinity

It is also clear that Paul Bates, like many other male professional staff at City Libraries, positively enjoyed his involvement in the male-dominated, 'technical' aspects of the job, an enjoyment and informal power position which was bolstered by his involvement in the exclusively male 'techies club', which often congregated out of office hours. This connection renewed and reinforced relationships between many men in the libraries department in a highly specific and gendered form. The following extract conveys the mixture of excitement generated through particular types of 'computer play':

> ... one was switched on, the flashing cursor, so it's positively inviting you to put something down. So I just put down 'menu' and a menu came out, 'Oh, great!', and I just started playing with it ...

The man concerned, a library research officer, went on to describe both his own interest in the technology, and more importantly, a 'gender-aware' account of the shared understandings which develop between those (men) who regularly 'play' with the technology.

> It's a bit more complex ... I think that those men that do [that], do it because they have no inhibition about being that sort of man ... and the way you learn is to spend hours in front of a screen, and *they've got hours to spend in front of screens*!

Most women, however, including the library assistants, many of whom worked part-time and rushed from work to pick up children or engage in other 'domestic' tasks, did not have 'hours to spend in front of screens'. In a connected way, few women were motivated to indulge in this kind of techno-play; as Fiorella de Cindio and Carla Simone suggest (see Chapter 10), women tend to be more interested in the *uses* to which it can be put, than in the technology itself.

Despite the fact that this 'self-exclusion' from forms of computer play which fosters formal skills and shared understandings clearly marginalizes women's positions at both formal and informal levels, it could be viewed as both a rational response, and/or a form of resistance to the highly gendered culture which surrounds such activities, which is a form of resistance also practised by men alienated from such 'masculine' cultures. As Suchman puts it:

> It becomes clear . . . that technologies and the work of which they are a part are not intrinsically gendered, but that relations of power and the meaning of technology are mutually constituted. This insight implies a re-framing of women's reluctance to engage in computer-related activity from failure to develop an interest, fear, lack of confidence and the like, to rational response and even, in many cases to a form of resistance. (Suchman, 1991)

An informal (self) exclusion referred to in the following extract from an interview between Mo Smart, the chair of the Design Team, and one of the (female) researchers, also confirms the existence of the 'techies club'.

EG: Have you noticed particular colleagues becoming computer buffs? Enthusiasts, you know, people who get the computer bug as it's called, then want to spend all their spare time on it, alone.
MS: We've got a few around.
EG: As I say, any in particular you want to name?
MS: No, I wouldn't want to name them, but there are people who play around: games at lunch time, golf courses on the screen and . . .
EG: Are they men rather than women?
MS: Oh, yes!
EG: Right [laughs] any women among them, ever?
MS: No.

Although the 'techies club' is obviously a very informal network, probably with a floating 'membership' of men who are not conscious of themselves as part of a group necessarily, the important point to note is the exclusively male nature of 'membership'. Also as the library research officer suggests, this activity reinforces particular kinds of masculinities through shared understandings and the creation of what could be considered an elite language, which contributes towards the maintenance and reproduction of a traditionally male power-base. In his own words:

> And once you've done that [spent hours in front of screens] . . . you'd know there were things you could exchange, a level of hints and tips, but more deeply a level of understanding, shared language . . . but they [those

> men] don't look around and see the absence of women, they don't perhaps think that they are creating a language, and perhaps, as it's expressed at work, a power-base which is exclusive, And they don't realize that they are where they are because the computing industry is designed for them. It's designed for people who *have* got hours to spare: in the garage, or the shed, or the attic, cracking the code.
>
> (See also Fergus Murray's discussion of the relationship between masculinities and technology, Chapter 4, this volume).

The fact that none of the women staff at City Libraries were part of this informal network is obviously significant in relation to their attitudes towards, and their level of confidence with the systems design process. It is also significantly related to the issue of equal opportunities in the organization. Women library assistants, particularly those working part-time were probably the least likely members of the 'techies club', mainly due to the heavy 'dual-role' responsibilities which part-time women workers assume (Sharpe, 1984). In addition, they have the most to gain at a formal level, and are the most in need of adequate implementation of equal opportunities policies.

Equal Opportunities Initiatives

As we outlined at the beginning of this chapter, during the 1980s Northfield Council had been committed to a range of radical policies, including the promotion of equal opportunities. Penny Cook, when appointed as City Libraries Director in 1983, had been able to develop a number of new initiatives within this framework, including the opening up of the career structure for library assistants.

However, in some respects the City Libraries experience illustrates limitations which have characterized many equal opportunities initiatives which focus primarily on pay levels and on recruitment, rather than on internal organizational change (Webb and Liff, 1988). Implementing equal opportunities policy in relation to gender inequalities also raises issues concerning the structure of hours, patterns of working, maternity and paternity cover and so on. All of these areas need to be addressed in clear consultation with the workers and trade unions involved. These complex issues emerged in particular in connection with the ways in which levels of library assistant participation on the Design Team became lower as the project progressed.

Like many clerical jobs, the library assistant role has evolved around the provision of various support functions and services; as Davies and Rosser (1985) have pointed out this is a model with strong gender connotations. In this kind of work, there is little or no time for reflection, for planning or for taking part in lengthy meetings with colleagues. These activities are more typical of professional posts. Participation in systems design also requires precisely these features, among others. Opening up such involvement to women clerical workers thus offers genuine opportunities for personal and career development, but without necessarily allowing the concomitant time and resources. Sustaining clerical involvement, therefore, really places on the agenda the issues of clerical job design and the interrelations between clerical and administrative or professional posts.

When problems surfaced within the Design Team, our own approach was to facilitate their emergence, and the ensuing process of negotiation. However, on many occasions issues did not surface explicitly enough to be resolved; library assistants felt too vulnerable to express some of their criticisms directly. Managers, in turn, felt both under pressure to deliver, in this new and uncertain context, and defensive when they sensed implicit criticism among Design Team members. In short, whereas the study circles made a definite break with past forms of IT-related activity in City Libraries, the Design Team did not achieve this degree of cultural change. By the time of our withdrawal from active collaboration, gendered patterns had been challenged in some respects, but confirmed in others. For instance, library assistants were now generally acknowledged as taking an active role in relation to future computerization, and their representation in groups and activities related to training and job design had been thoroughly consolidated. When a 'technical' group was formed as part of the project reorganization in July 1991, however, it was composed entirely of male professional staff.[3]

Two Sets of Unresolved Tensions

Reflecting upon the City Libraries research, we can identify two sets of unresolved tensions, both of which have implications for human-centred systems strategies and methods.

First, the case-study illustrates a paradox, concerning systems development and user relations. User-involvement is increasingly important to the successful design and implementation of complex systems. At the same time, it brings to the surface issues which cannot be tackled successfully within the confines of any single 'project' or indeed by the HCS approach, which fails to incorporate a modern division of labour (Hales, 1992) and therefore cannot adequately theorize the (gendered) relationships between 'users' (for example, library assistants) and managers. Thus, large-scale processes of systems development represent major opportunities for reviewing and changing established patterns and practices in organizations. Inevitably, however, conflicts and uncertainties are also revealed, and responding to these requires approaches which go far beyond conventional systems development methodologies and techniques.

In City Libraries, for example, library assistant participation clearly strengthened the systems specification and evaluation processes, enabling potential implementation problems to be anticipated and avoided. However, this participation also exposed more clearly than ever before the conflicting demands which changing library policies were already placing upon the library assistant role. That is, the day-to-day pressures to deliver a 'service', in a historically feminized form, were intensifying, in a context of cutbacks in public expenditure. At the same time, service delivery in general, and systems development in particular, required new levels of collaborative decision-making, reflection and planning, which had historically been excluded from the library assistant role. The *de facto* alliance with the City Libraries director, which underpinned the project when it started, suggested a degree of scope for gendered patterns in library assistant career and job structure to be explored and challenged. However, the later combination of cutbacks, and a more cautious senior management style, undermined these possibilities.

Second, the City Libraries case-study raises issues concerning the relationship between social science and computer science perspectives on systems development. Social science analyses can contribute to the facilitation and review processes which sustained user-involvement requires, but a basis of genuine exchange and partnership is required, rather than two perspectives or contributions which remain separate or parallel (Ehn, 1988). Achieving this exchange in reality represents a considerable challenge to established practices and definitions. It necessarily requires a form of self-reflexivity or self-awareness of one's own context and experience, as both theorist and practitioner, since a precondition of working across discipline boundaries involves an understanding of the strengths and weaknesses of various theoretical perspectives and empirical approaches. Our case-studies have spanned an unusually lengthy period in research project terms, which afforded us ample time (in the early stages) to engage in lengthy discussions on the merits of different theoretical approaches. However, the complexity of the action research involved in being part of an ongoing, radical systems design process, resulted in the need for a major coordinating role, and this required almost unlimited time and energy. Nonetheless, our own understandings and definitions of both human-centredness and gender perspectives shifted considerably during this period and not always in a shared direction! (See the Introduction to this volume, and Chapters 1 and 2 for an overview of developments in computer science, social science and perspectives on gender).

HCS Strategies and Gender Relations

Reflecting on our case-study account and on the two areas of complexity outlined above leads us to propose a number of general points or guidelines in relation to HCS practices, complementing the general principles proposed in the first chapter.

Aims

First, we would propose the following aims for interdisciplinary initiatives in HCS research and practice:

1 To prompt critical debate, and new practices, concerning the design and use of office information technologies;
2 To support the articulation and negotiation of different interests and perspectives, within organizations, concerning information technology;
3 To identify opportunities for intervention and change in connection with the patterns of gender inequality which are currently associated with information technology design and use;
4 To develop innovative practices in the design of systems and of working practices.

New links between IT strategies and equal opportunities strategies

Second, we recognize the severe limitations of the ways in which legislation, policy and political debate have all been framed with reference to 'equal opportunities'. However, equal opportunities structures now exist in many organizations; in some instances, they coexist with more far-reaching initiatives on gender. We would suggest that both can form a point of departure for challenging the conventional boundaries between strategies on IT and those on other aspects of policy development. Therefore:

1 At an organizational level, both equal opportunities strategies (or related initiatives on gender and other inequalities) and IT strategies should be strengthened by the establishment of links and cross-references, broadening their scope.
2 These interconnections should be established, first, by ensuring that policy groups or committees in each area each include at least one person responsible for liaison with the other; second, by ensuring that policy documents and proposals address areas of common interest such as training and job design in which joint proposals can be developed.

Methods and techniques

Third, the study circle and design team initiatives described here do not supersede existing methods in systems development — such as prototyping — which can offer opportunities for user-involvement. They do, however, illustrate ways in which a strong basis can be created for women clerical workers to collaborate with designers and with other systems users. In that sense they can be described as relevant elements *within* a human-centred methodology which would also encompass techniques such as prototyping.

Study circles are particularly relevant to the processes of identifying problems and opportunities associated with office systems development, and of encouraging user-involvement on a broad basis.

Study circles:

— should take place in work time;
— should not include more than ten participants, of a broadly similar status;
— should have the resources both to assess examples of new office systems in use outside the organization, and to examine existing practices within the organization;
— need a facilitator, rather that a leader or trainer; this does not have to be someone with computing expertise, or someone outside the organization.

Study circle groups should also have the scope to report their observations to senior management, to trade union representatives, and to other relevant policy bodies in the organization (such as an equal opportunities or IT committee).

A design team structure should include representation from every level within the organization; clerical representation should not be less than one third. Clear lines of communication with management and trade union bodies are essential, as is a clearly-defined convenor or chairperson. The brief of the team should be to address 'systems development' in a broad sense: that is, to identify issues of policy and of organizational change, as well as identifying IT possibilities.

A range of techniques are appropriate, to support the work of study circles and design teams; they need to include, but are not restricted to, the following:

1 Techniques for visualizing new possibilities in connection with IT, such as paper-based mock-ups; computer-based mock-ups and prototypes; practical assessments of systems in use; systems demonstrations;
2 Techniques for specifying the design both of software and of working practices, for example, formal specification documents; samples of screen design (on paper and computer-based); model job descriptions;

3 Techniques for evaluating systems, and working practices, both before changes are planned and after implementation, for example, questionnaires, interviews and informal group discussions.

Conclusion

The original aim of HCS research was to embody values of collectivity and solidarity in new designs of working practices and of technologies. However, the limits of a model built upon an inadequately theorized concept of social class are increasingly obvious; as Hales (1992) argues, 'humans' referred to 'ordinary people' which in practice meant 'somebody who is one of Us, not one of Them' — not a '*Boss*' (p. 12) — This concept obscures the complexity of social relations in workplaces such as that described in City Libraries, where the professional-managerial strata are both the decision-makers and the (often supportive) colleagues of junior clerical workers like the library assistants. Feminist accounts of gender relations in the workplace deal more adequately with such complex hierarchies of relationships. (see for instance Chapters 6 and 7 this volume). It may be that as Hales suggests[4] the emergent approaches of Computer-Supported Cooperative Work (CSCW) can encompass gender perspectives and offer (among others) a more fruitful theoretical framework through which to explore the issues discussed in this chapter. Meanwhile, the case-study observations discussed here underline the importance of continued research and debate in order to consolidate and expand a range of gender perspectives in relation to office systems development.

Notes

1 See Hales (1992) for a detailed discussion of the evolution of (British) Human-Centred (HC) approaches, both theoretical and political.
2 The Human-Centred Office System Project based at Sheffield City Polytechnic incorporated two phases of development which were (separately) funded by the UK, ESRC/SERC Research Councils Joint Committee on the Successful Management of Technological Change. Phase 1 (1984–87) and Phase 2 (1987–91) comprised different, although linked, research projects and research teams. The findings discussed in this chapter relate to Phase 2 of the research.
3 At the time of writing, City Libraries' Design Team and Departmental Management Team had invited formal tenders from systems suppliers. However, continuing financial crisis within the local authority prevented a final implementation date from being agreed.
4 See Hales (1992) for a detailed analysis of the potential merits of CSCW as an alternative set of conceptual and empirical approaches to those of Human-Centred System models.

References

Beechey, V. (1987) *Unequal Work*, London, Verso.
Bijker, W., Hughes, P. and Pinch, T. (1987) *The Social Construction of Technological Systems*, Cambridge, MA, MIT Press.

BRAVERMAN, H. (1974) *Labour and Monopoly Capital*, New York, Monthly Review.

COCKBURN, C. (1985) *Machinery of Dominance: Men and Technical Know-How*, London, Pluto.

COOLEY, M. (1987) *Architect or Bee? The Human Role of Technologies*, London, Hogarth Press.

DAVIDSON, M.J. and COOPER, C.L. (1992) *Shattering the Glass Ceiling: The Woman Manager*, London, Paul Chapman Publishing Ltd.

DAVIES, C. and ROSSER, J. (1985) 'Gendered Jobs in the Health Service: A Problem for Labour Process Analysis', in KNIGHTS, D. and WILLMOTT, H. (Eds) *Gender and the Labour Process*, Aldershot, Gower

EHN, P. (1988) *Work-Oriented Design of Computer Artefacts*, Stockholm, Arbetslivscentrum.

ERIKSSON, I.V., KITCHENHAM, B.A. and TIJDENS, K.G. (Eds) (1991) *Women, Work and Computerization: Understanding and Overcoming Bias in Work and Education*, Amsterdam, Elsevier/North Holland.

FRIEDMAN, A. and CORNFORD, D. (1989) *Computer Systems Development: History, Organisation and Implementation*, London, Wiley.

GASKELL, J. (1987) 'Conceptions of Skill and the Work of Women', In HAMILTON, R. and BARRETT, M. (Eds) *The Politics of Diversity*, London, Verso.

GILL, K.S. (1990) 'Summary of Human-Centred Systems Research in Europe', Working Paper, Seake Centre, Brighton Polytechnic.

GREEN, E., OWEN, J. and PAIN, D. (forthcoming) 'Designing Systems, Defining Roles' in WOOLGAR, S. and MURRAY, F. (Eds) *Social Perspectives on Software*, Cambridge, MA, MIT Press.

HALES, M. (1992) 'What Has Gender Got to Do With Human Centred Design? or Why Don't we do more things different?' (Draft) Paper for the CIRCIT, Melbourne, Australia, August 1992.

HARAWAY, D. (1991) *Simians, Cyborgs and Women — The Reinvention of Nature*, London, Free Association Books.

HARDING, S. (1986) *The Science Question in Feminism*, Milton Keynes, Open University Press.

HIRSCHHEIM, R. (1985) *Office Automation: A Social and Organisational Perspective*, Chichester, Wiley.

LIBRARY ASSOCIATION (1989) '*Equal Opportunities Monitored*', (Internal Report).

MUMFORD, E. (1987) 'Sociotechnical Systems Design — Evolving Theory and Practice' in BJERKNES, G. *et al.* (Eds) *Computers and Democracy — A Scandinavian Challenge*, Amsterdam, North-Holland.

MURRAY, F. and WOOLGAR, S. (1990) 'Social Perspectives On Software', Report produced for the ESRC, '*Programme on Information and Communication Technologies*' (PICT).

OLERUP, A., SCHNEIDER, L. and MONOD, E. (Eds) (1985) *Women, Work and Computerization: Opportunities and Disadvantages*, Amsterdam, North-Holland.

OWEN, J. (1992) 'Women Developing Office Systems: Case-Studies in User-Involvement During Design' PhD Thesis, Sheffield City Polytechnic.

ROSE, H. (1983) 'Hand, Brain and Heart: A Feminist Epistemology for the Natural Sciences', *Signs*, **9**, 1, pp. 73–90.

ROSENBROCK, H. (1989) *Designing Human-Centred Technology: A Cross-Disciplinary Project in Computer-Aided Manufacture*, Amsterdam, Springer-Verlag.

SHARPE, S. (1984) *Double Identity: The Lives of Working Mothers*, Harmondsworth, Middlesex, Penguin.

SILVERMAN, D. (1985) *Qualitative Methodology and Sociology*, Aldershot, Gower.

STONE, I. (1988) *Equal Opportunities in Local Authorities: Developing Effective Strategies for the Implementation of Policies for Women*, London, HMSO.

SUCHMAN, L. and JORDAN, B. (1989) 'Computerization and Women's Knowledge', in

TIJDENS, K.G. *et al.* (Eds) *Women, Work and Computerisation: Forming New Alliances*, Amsterdam, North-Holland.

SUCHMAN, L. (1991) Closing Remarks on the 4th Conference on Women, Work and Computerization: Identities and Differences, in ERIKSSON, I.V., KITCHENHAM, B.A. and TIJDENS, K. (Eds) *Women, Work and Computerization: Understanding and Overcoming Bias in Work and Education*, Amsterdam, Elsevier/North Holland.

VEHVILAINEN, M. (1986) 'A Study Circle Approach as a Method for Women to Develop their Work and Computer Systems', Paper to the IFIP Conference, *Women, Work and Computerization*, Dublin, August.

VEHVILAINEN, M. (1991) 'Gender in Information Systems Development — A Women Office Workers' Standpoint' in ERIKSSON, I.V. *et al.* (Eds) *Women, Work and Computerization*, Amsterdam, Elsevier/North Holland.

WEBB, J. and LIFF, S. (1988) 'Play the White Man: The Social Construction of Fairness and Competition in Equal Opportunities Policies', *Sociological Review*, **36**, 3.

WEBSTER, J. (1990) *Office Automation*, Hemel Hempstead, Harvester Wheatsheaf.

WEST, J. (1990) 'Gender and the Labour Process', in KNIGHTS, D. and WILLMOTT, H. (Eds) *Labour Process Theory*, Aldershot, Gower.

WILLIAMS, R. (1990) 'Participation and New Technology: Theoretical Framework and Research Hypotheses in Attitudinal Survey', Working Paper EF/WP/90/30/EN European Foundation for the Improvement of Living and Working Conditions, Dublin.

Chapter 9

Strengths and Weaknesses of Participation: Learning by Doing in Local Government

Mike Hales and Peter O'Hara

This chapter draws lessons from five years' work on a major computer system design project. The project continues; the chapter covers the formative period from 1986. The project is within the social services department (SSD) of a London Borough — a large English local government authority. The central function of the computer system is to hold records of SSD's dealings with clients. This is important because multi-agency practices and related accounting procedures have become the norm in social services delivery.

Local government staff are divided into clerical–administrative and professional grades. Both groups are users of the SSD system; professionals generate much of the data in their consultations with clients. Initially there were no computers in SSD. By 1991, there was a wide area network with three decentralized mini computers and eighty-five personal computer workstations; with a planned increase to 150 by 1993. The project is currently in implementation Phase Two, focused on the second major subsystem of the system architecture. It has been led by one of the authors as an SSD middle-manager, required to deliver an installed system within real time and budget constraints, in an organization whose environment was changing dramatically. The project has been strikingly successful in these terms, contrary to previous experience in the department.

However, this report is not simply a success story. The project started out in a highly political setting, with ambitious and unusual goals. It has been framed by an ideology of participation and bottom-up activism. We will discuss this environment in relation to issues of organizational know-how and managerial competence in a large bureaucratic organization. We will explore the project's gender content and discuss whether the gender outcomes at this stage are different than might have been expected from a more conventional development approach.

This evaluation is incomplete, and we want to stress that this report is essentially about learning in a live setting. First, the methodological framework is patchy. Actors in the project have invented methods and interpretations as they went along, informed by an overall, relatively explicit sense of what kind of (social) development was wanted; they have referred to as much (system development) precedent as was available to them in the project setting.

Second, the gender analysis is patchy. Much of the practice was framed by 'equal opportunities' as distinct from 'gender', and we are dissatisfied because we are still unable to write the account that we would like to write, in terms of gendered practices and empowered action as distinct from the cultural politics of bureaucratic guerilla warfare: missions, budgets, departments and top-down project control. These two modes of analysis sit awkwardly together in what follows, but we feel that it is important to show them in this unhappy partnership.

Finally, there are many systematic connections which ought to be made with what is now a rapidly growing literature on 'computer-supported cooperative work' (CSCW), 'organizational computing' and related threads of information systems research. However, when we started to write this chapter we were unaware of much of this literature (apart from Scandinavian approaches to collaborative design) and we feel it may be more helpful to try to describe the project as it happened rather than to reconstruct it in terms of a currently improving academic understanding of computer systems, workgroups and design as a social-cultural activity.

The remainder of this paper has seven main sections:

1 The context — why was a participatory approach supported?
2 Participation — 'circles' as a practical form for design;
3 Technique in support of participation — prototyping;
4 Participatory challenges of implementation and use;
5 Empowerment, management and participatory ideology;
6 Equal-opportunity practice — plans and situated actions;
7 Conclusions — politics, design and management.

The final section touches on some knotty issues in design politics: what does the term 'human-centred' apply to? Are front-line staff in SSD more empowered and in control now than before? What is the relationship between academic research and learning by doing? Could more capable management mean that equal opportunity objectives might be equally well met in the private sector as in some 'radical' public-sector settings? And to what extent would theoretically-explicit gender-awareness matter, if some other modes of empowerment were more active than they currently are?

The Context — Why Was a Participatory Approach Supported?

In the 1980s local government in the UK went through large changes because of central government legislation. The SSD project is set in a highly volatile period. Previously stable power relations were shifting — first between SSD and the central information technology (IT) group in the council (which historically had taken the lead in all computing developments) and second in the attitudes and responses of staff and mangers in SSD.

SSD had a computer liaison team (CLT); the head had some previous IT-user/manager experience; his female colleague was trained as a business analyst. These two jointly set the tone for IT issues in SSD; both were also active trade union members. They promoted the idea with senior managers that it is both right and possible to redefine IT strategy and design issues in non-technical, client-led, general management terms (service quality and content, managerial control, culture and style, equalization of opportunities). This ideological lead from semi-technical, lower-middle professionals was crucial in shaping the project.

In the council, as a large bureaucratic organization, there was the typical power struggle between user departments and the central IT department, rooted in an applications backlog and repeated failure to deliver departmental systems. In early stages, participatory issues were foremost in CLT's agenda — involvement of eventual users in design. The council's IT group failed to recognize or respond to this 'unusual' request. Two senior SSD managers, supported by CLT, took leading roles in an ensuing political battle. The initial contested terrain was IT budgets and their control (demands for a client-led project). This expanded to include technical issues: IT standards, the IT project management approach, development methods, criteria for evaluating tenders and designs. The council's IT professionals tried to represent SSD as unprofessional, a danger to the council's established systems and procedures. Eventually SSD managers won the support of the controlling committees for a client-led specification, tendering and design process; the internal contractor was excluded from bidding for the work.

Central government legislation had introduced compulsory contracting and tendering (CCT) into some areas of local government practice. The 'market' values of CCT had caused cultural upheaval in the council. This was aggravated in SSD by further legislation on community care and care of children. Difficult questions were posed: what services should be delivered and how; what constitutes professional practice in this new era? At the heart were practical questions of financial management, resource management via multiple agencies, and formal processes for securing accountability. Managers in SSD — typically, in local government — have generally been promoted from professional social-work positions, with little managerial training.

There were emergent differences between the values and self images of managers, professionals and staff. Especially, there was a gap between staff involved in front-line service delivery (mainly women) and others responsible for strategic decisions (mainly men). CLT, typically throughout the project, aligned itself with front-line views and values, while at the same time supporting the notion that top-down decision-making practices must be made rigorous, implementable and accountable. With hindsight, it could be said that this combination of commitments — 'political' (front-line, serve-the-people, bottom-up) and 'professional-managerial' (effective top-down management; effective administration) — put CLT into a rather small minority in SSD.

Equal opportunity issues — especially women's employment status and opportunities — were important and explicit from an early stage. Formal equal opportunity policies were ideologically important in the council, and the values were well entrenched in the department's culture. Members of CLT were personally committed to them, and saw the project as a significant locus for equal opportunity initiatives, via training, job design and changes in the status and awareness of participants. At the outset both of the senior SSD managers involved in the inter-departmental power play were black, one a woman. She was closer to the project and thus, in negotiations with the IT department, had to operate in a white, male, techie sphere. Staff who were to participate at various stages — the final users of the system — were predominantly female. These conditions sharpened the early awareness of gender issues.

A Space, Recognized and Occupied

Why was a human-centred, participatory, gender-aware approach to developing an integrated business system sanctioned? Answer: a space was opened up, and it was recognized, developed and occupied in a particular way by key actors.

The project emerged from a major review, in 1985, of SSD's approach to and experience with computer systems. Previous attempts had produced few tangible benefits, but by the mid-1980s both need and demand for modern systems were seen to be critical to SSD's performance and development. The opportunity to explore a more progressive strategy arose from this perception and the inability of the in-house IT contractors to deliver urgently needed IT systems, forcing the department towards a client-led approach. The space was enlarged by the failure of in-house IT professionals to recognize the department's wish for a design approach involving final users. In the battle between client side and contractor side for budgetary control, the political 'space' took the form of a high-risk, high-profile, client-led process of specification, tendering and design.

This space was recognized, developed and occupied by actors (CLT) within the social services culture, who accepted and promoted SSD's participatory ethos. The key role of change agent and champion fell to the team; most of CLT's influence grew by expanding into the vacuum created by lack of relevant know-how among senior SSD managers. These managers saw CLT as trusted 'friends', possessing technical skills (or at least, not frightened of techies) and an alternative, plausible, acceptable vision of IT development. If these judgments have been subsequently validated, it was more by luck than strategy on the part of senior management.

The gender dimension of human-centred design was introduced by these key CLT actors, and supported by senior managers early in the project. A rhetoric of human-centred, gender-aware, bottom-up development was consistent with SSD culture, and helped to unite the department in its struggle with unsympathetic techie outsiders.

Participation — 'Circles' as a Practical Form for Design

In germinating the project, CLT had to both engage the organizational agendas and develop a methodology which would underpin the project's unusual (and in this organization, unprecedented) approach. With 'participation in design' — for both managers and end users — as the generating methodological notion, design teams were established in the spring of 1986. Being within the flow of a live project, design teams cannot be entirely open ended and self determining. However, the aim of CLT was to approach these conditions as nearly as possible within the constraints of 'project time' and the project-manager role. Several principles distinguished the management of design teams in this project, compared with conventional approaches to user involvement:

— 'vertical' groups with majority membership of *front-line final users*; not just managers, professionals and computer professionals;
— voluntary part-time membership with *positive discrimination* so that numbers of women and black workers reflected the composition of the user constituency;

—a focus on *personal empowerment* of individuals, so that contributions by all participants could be equally registered and valued. Managers and techies were not allowed to dominate; their contributions were required to be intelligible and acceptable to other members;

—a *facilitator* was used to support group development; teamwork exercises were part of the design-team curriculum;

—*mutual learning* was a cornerstone; design teams were seen as knowledge-producing workgroups, with a partially self-determined 'curriculum', requiring support with educative, informative and research inputs (briefings, visits, etc.);

—forming a '*whole subject*'; producing adequate design knowledge of a complex Whole Object (that is, a major departmental computer system) requires a Whole Subject. Such a Subject is unlikely to exist 'naturally' within an organization, and must therefore be deliberately organized into existence, crossing conventional boundaries and creating a new domain of conscious practical activity;

—*line management* as a separate domain; design teams are knowledge-producing, not formal decision-making groups. Designated managers remain accountable (both upwards, to their seniors, and downwards, to the group and its implicit constituencies) for the use they make of the teams' work.

CLT's members were the main facilitators. However, these people were inexperienced, and skilled outsiders were brought in at each stage to augment the team's activities. CLT members were not very high in the departmental hierarchy and therefore, as facilitators, did not carry too much formal role-power.

Design teams organized on the above lines might better be called design *circles*, to underline the differences with conventional design organization. 'Circles' comes from 'study circles', as used in Scandinavian approaches to technology design, especially with women (Vehvilainen, 1986; Greenbaum and Kyng, 1991; Ehn, 1989). Study circles are open-ended educative groups, whose learning and research agenda is self-determined. They are groups of technology users and potential users, whose members are given equal status within the group, and in particular are all regarded as bringing relevant knowledge to the discussion of technology in the context of its use. CLT were aware of study circles through academic contacts.

Design teams had different functions in each of four project stages: tendering, strategy study, Phase One development, Phase Two development.

Tendering: three teams (vertical, within key work areas) were formed to develop the requirements spec. In addition to technical factors, the spec included organizational and 'style' conditions (for example, participatory, equality-aware practice). CLT evaluated bids according to the spec, and shortlisted them. Then the teams plus senior management evaluated these bids, further reduced the list and made the final selection for presentations. Team members compiled an interview checklist before attending suppliers' presentations — a tough and unusual situation for suppliers. One technically excellent supplier was marked down for presenting a computer dating system as their model — completely out of key with the anti-sexist local ideology. Through this process, final end users, as members of design teams, were directly involved in choosing the software development environment and the contractor.

Strategy study: the successful bidder (a 4GL company) carried out a top level business strategy study. This was a key factor in their winning the contract. The study involved higher management and CLT, excluding staff lower down the hierarchy (such as the design teams). It produced an analysis of both system architecture and organizational practice, leading to both a proposed system design/implementation programme *and* a proposed programme of top-down organizational/cultural changes. The study was thorough but produced an impossibly long, complex and jargon-bound report, so a translation was commissioned (Hales, O'Hara and Smith, 1988). This served as a management briefing document, setting out ground rules and interpretations for the remainder of the project: entity and function models, architecture, core implementation plan, further design decisions required.

Phase One development (detail design and the first system implementation): the first system was a core system for the whole of the department. Custom overlays, with additional detail, tailor it to local use. Two design teams were formed — a 'focus team' in an application area, and a generic 'validation team'. Prototyping began, with design teams as the evaluators. In addition a 'strategic design team' was created. Composed entirely of senior managers, this was intended to be the sign-off body, making final design decisions and progressing the contract. After a couple of meetings, however, it failed to reconvene.

Phase Two development: detailed design of the second core system started while Phase One was being implemented. Previous success and acceptance of the participatory design principle made additional resources available (for example, increased training support). But staff cover for attendance at design team meetings was still not provided. There were three design teams, two of them generic and multi-disciplinary.

An Illustration of Design Politics

One of the Phase Two design teams, a single-discipline group, hit problems. An explicit aim of the system architecture (an output from the strategy study) was to support reforms in professional practice. Thus the architecture mapped a reality different from that recognized in current professional standards, custom and practice. The hardware/architecture platform had been designed for evolutionary implementation, so local practices may in some respects change slowly within the framework of the project plan, giving space for local politics to be resolved. In this part of the organization there was a major divergence of views and interests between administrators and social-work professionals. Professionals wanted to tie admin staff to computers, whereas the architecture required professionals to become routine computer users — the integrity of the data (and thus the quality of the support provided by the database) depends on it. This conflict is probably strongly gender determined, and is not yet resolved.

The Shifting Balance between Women and Men

The historically dominant technical professional group in the council (IT department) was marginalized by the client-led approach, by the success of an outside contractor in the bidding process, and by the participatory form of design practice. In design teams, external power structures were suspended as far as possible by the facilitator's deliberate actions; managers and techies were represented rather than in control. The contractor's principal analyst and technical project manager were women. All the facilitator/trainers employed were women (a deliberate choice by CLT). As a result (we assume) of these conditions, female members of design teams do feel that they were given an unusual quality and quantity of space during design.

At two stages (end of the pre-tendering phase; during Phase Two) external researchers were brought in to evaluate aspects of the project, including the extent to which design-team members felt involved in and empowered by the participatory process. Both studies indicated that in these terms — that is, as a participatory system design project — the project was succeeding. However, this only serves to highlight the limits of a 'system design' perspective. Most of the major benefits for female staff still lie in the uncertain future, when the system is fully implemented and *in use*, with revised divisions of labour, career paths and work roles, across the whole organization.

Although social services has a female majority of staff, there is a male majority of managers. Early on, CLT was careful to ensure that design teams had a female majority. In the implementation stages, however, men have figured more because a) as the project became visible, with workstations in place, men have moved in to acquire skills and attach themselves to its status; and b) middle management is the critical area now, and managers in this organization are male. Although CLT created structures that would give women a space in which to develop different modes of action (for example, as designers or researchers), once the project became a recognized part of the political scene, male managers acquired space and the balance shifted, with full consequences yet to be identified. The project leaders had not anticipated this.

By 1991, changes had taken place in staffing and in the departmental power balance. The CLT head left, with no immediate replacement; the IT department came back into the project, filling the vacuum thus created. Therefore the approach developed during the first four years also shifted, in favour of a conventional professional-managerial approach to implementation. Consequently, the original perception of the project situation, as an opportunity to realize significant, widespread changes in the male/female balance of skills, ownership and access to technology, began to recede.

Technique in Support of Participation — Prototyping

It needs to be said that, even with today's technology, prototyping is not easy. This project was a learning experience for both SSD and the system suppliers, and even though both parties now have more experience, the necessary learning in each and every future case should not be minimized. With hindsight, we see the benefits of prototyping as:

- —enabling techies to understand enough of SSD practice to deliver a useful design;
- —enabling techies to make a technically-based contribution to the reflection of future users, via the prototype-Object itself, and a technically-based commentary on it; and
- —enabling users literally to see what their work might begin to look like, rather than to talk in abstract terms.

That is, prototyping is a) a communication setting, created to mediate between users and technical designers, and b) a visualization technique for non-technical designers.

Because there was a need to explore the behaviour of the system as a whole, in Phase One the prototype was developed almost into a full, workable form. Then the prototype was scrapped, and the working system designed from scratch. That is, prototyping was used by technical designers as a *knowledge-extraction technique* or specification technique, preceding more abstract, formalized analysis and design by computer specialists. Prototyping in Phase One took longer and had more iterations. CLT and technical designers had a dual task: learning *how to create a practice* for exchanging and producing complex knowledge, and also learning how to *recognize and represent* the structures, processes and behaviours of this specific design object.

In Phase Two the design focus was more superficial: essentially, the interface — screen design. Organizationally, there was less to learn at this later stage, but the prototype itself was equally complex, because the (more limited) design object was addressed in more detail. In a subsequent project by the same partners, Phase One would be simpler, and possibly quicker. Learning '*how* to do it' takes the most time, as distinct from *doing* it.

Communication and the Machine

In detail design stages (Phase Two onwards) there is a (real, not prototype) core system to show, which can be used for training design team members. Before being exposed to techies, Phase Two design team members were given separate training to familiarize them with 'the system so far . . .' In Phase One, however, the techies brought the first prototype as a 'strange object' to the other members of the design team. This shifted the ground, and presented the possibility that power (the ability to 'talk sense' about the Object of discussion) might shift from users to techies at this point.

This problem is not insuperable, but it is real. On the machine, a draft interface can be too opaque and 'monoptic' (even if multiple windows are used), too predetermined and closed. It is not clear whether this is a) an inevitable factor in technofear among inexperienced users, b) a passing characteristic of the present stage of 4GL and graphic interface technology, or c) an intrinsic limitation of the virtual-machine medium. These were major concerns during Phase One. In Phase Two, the techies knew enough about the users to be able to open the discussion in ordinary (SSD) language rather than simulation language.

Importantly, prototyping in Phase Two was carried out at first using hand-drawn representations of screens, on paper. The next time around, Phase One would use paper too, at first; this change requires more analysis to be carried out beforehand. Using a machine can cause unnecessary problems.

There is a sidelight on diagramming. Some people are turned on by 'hands on' experience with machines (and others are paralyzed); some by talk; others by diagrams. With foresight, diagramming, prototyping (both paper and machine-based) and other alternatives can be 'scheduled' to give a good communications mix. Diagramming is helpful to techies, giving them permission to use professional skills (entity diagrams, data flow diagrams, etc.) in the design circle setting. Helpful insights into these practical tactics are found in Part II of Greenbaum and Kyng (1991).

Participatory Challenges of Implementation and Use

In Phase Two some problems emerged. Support available from CLT was more limited (staff vacancies were not quickly filled), but by this time the contractor's staff understood the style quite well, and no special problems arose with user participation in design teams. However, by then the project involved a wide range of people, and there was a sense that the political vision, the mission of empowerment and managerial effectiveness, was not as alive as it should be. The process became institutionalized. Participation dried up in the implementation phases — developing stopped, and pre-designed standard systems were being inserted routinely into workplaces.

The organization — notably senior and middle management — did not yet thoroughly own either the architecture or the service/flexibility/enabling/empowering principles embodied in it. Hardly any management development activity followed the strategy study and the 'Progress by Design' briefing document. Thus there is still a lack of strategic management awareness, regarding the design of both jobs and broader practices in relation to the opportunities created via the 'system agenda'. To aggravate this fundamentally difficult situation, the animating energy of CLT, as guardian of the project's vision, has become bogged down in the delivery responsibilities of a large system-design and implementation programme.

A crucial failure flows from this: a whole dimension of human-centred design has been lost in practice, even though it was recognized as a need by key actors. A participatory system-design practice has been designed and implemented, as distinct from a design/implementation/use practice. A complex, powerful and relevant system, valued by those who contributed to it, has been designed and developed in detail, as distinct from a system-use practice. Some central propositions of human-centred design (designing a technology is designing jobs; designing a system life cycle is designing careers, training regimes and equality of opportunities) have drifted into the background and are invisible to most of the people who now find the system in their workplaces. Since many of the job design and career issues were women-centred, gender awareness has (temporarily?) lost its main footholds in the project and opportunities are not being created and seized as originally intended.

So where do 'circles' and participation go from here, in implementation and use rather than design? Our view is, in principle, into a debate about the newly installed IT apparatus as a system of 'tools for conviviality' (Illich, 1973), an opportunity for individual and collective learning and advancement, and a setting for positive action over access to career and learning opportunities. There is a need to discuss:

— *allocation of marginal-cost resources*: positive discrimination as a criterion for allocating resources during implementation (training, standard software that will run on the installed hardware at low marginal cost);
— *'idiot-proofing' as an aesthetic of detail design*: who wants to be cast as an 'idiot'? Why is this bad? The architecture permits flexibility; so how much ongoing (re) design power should be left visible on the face of the system and its official practices?
— *operational support and routine user training*: how much to provide, where to locate it?
— '*constructive users*' (Hales, 1991a; Carroll *et al.*, 1989): how can they be identified, supported and helped to develop both personally and professionally, within a deliberately modified departmental division of labour, a formal system of vocational qualifications and an explicit career progression scheme for information workers?
— *evaluation and review*: how to review the design and manage entry into a 'design-in-use' and redesign cycle.

How to do this? A limited step would be to create technology skills drop-in workshops for users of the new system. With an open-ended skill-sharing format and the support of a trainer with some facilitator skills, some of the job and career design issues could be explored; workshops could function as study circles, at a technique-oriented level. A more powerful move would be to leave design circles with their decreasing domain of relevance (that is, the detail design of remaining subsystems) and branch out into quality circles as part of a programme of organizational development linked directly with system implementation. The name's different but the game's the same (or could be), with the important difference that the service rather than the (IT) system would now be the explicit agenda of a 'circle' programme.

This project moved straight into design teams, based on the study circle model. As the system moves towards its first review, study circle programmes will be called for. Study circles are important for redesigning and upgrading existing systems, where there is a base of current experience that can be systematically explored. There will almost certainly be a base of dissatisfaction and frustration, stemming from underdesign, undercapacity, poor implementation, poor job design and poor management. The open-ended, participant-centred, supportive, bottom-up form of study circles is necessary in approaching the typical tangle of inexplicit knowledges and conflicts and redressing the imbalance of top-down action.

Empowerment, Management and Participatory Ideology

The driving force of the project has been a commitment among key actors to empowering female staff in the context of information systems design and use. 'Participation', as a slogan, is a 1980s coded version of 1970s revolutionary socialist ideology — widely endorsed within social services professional work. However, there is a need for middle-level people, in staff rather than line positions in a large bureaucratic organization (such as members of CLT), to also work 'top-down'. The hierarchical structure of the organization makes this necessary if they are to reach other parts of the organization, creating spaces, providing resources and encouraging, recognizing and supporting self-aware action by staff at lower levels. To an extent the bottom-up ideology of participation has obscured this fact of political life in big organizations.

The practice of participatory design is reformist, and in day-to-day terms, other slogans such as 'accountability' and 'competence' also turn out to be important — not least because they refer to the situation of professionals and managers rather than the rank and file. Most of the failure that we recognize in the SSD project can be located in the top-down as distinct from the bottom-up dynamics. Managers were too frightened or incompetent to deliver the changes that staff lower down need, want and deserve. In the present circumstances of professional/managerial work in the local government sector, managers need empowering too. Too many of them cannot and will not take risks, even when their ideology (participation, equal opportunity, etc.) logically implies it. We feel that this general shortcoming of management is aggravated by (rather than rooted in) the specific factors of technofear and IT illiteracy.

Human-centredness (HC) is not just about utopian visions (or even visions which take the tangible form of a system architecture or a configuration of apparatus). It is about delivering real change in people's ability to act and to perceive possibilities. Participation — and the design circle, as a concrete form of participatory activity — is simply not a big enough concept to frame all the necessary action. HC in large hierarchical organizations has to be about top-down as well as bottom-up processes: responsibility, accountability and empowerment as well as participation and enabling; scale as well as quality; what managers produce as well as what users use. HC is about both learning and delivering; and in large formal organizations, both of these come home to roost with managers.

The Management of 'Design'

Some of the issues we address can be located in the dynamics of the design/development cycle. Design properly should be understood as the whole cycle of design/implementation/use/evaluation/redesign (Hales, 1991b). Within the cycle there are subcycles with both shorter and longer timescales (iteration during design, 'tuning' during use, etc.) and the phase normally called implementation itself includes aspects of design: 'design-of-use' (as distinct from design of a hardware/software system) and 'design-in-use' (as distinct from up-front design). Juggling the dynamics of such a complex, multi-process, multi-cyclical process is no easy matter, and SSD's project leaders were novices. Although they implicitly accepted the whole-cycle model of design, they were exploring its detailed practical implications step-by-step.

Thus, as it happens, a combination of external circumstances and poor decisions by CLT led to staff resources being very tight at the beginning of Phase One; the result of this, in turn, was that activity centred heavily on 'design' in the conventional, narrow sense. Implementation was starved of effort. It was during this phase that issues of managerial empowerment (the management of change) became more obvious, and in other circumstances — with other styles of project management — might have been better addressed.

Conventional project management practice would have frozen the design after Phase One and implemented the system in that partially-complete form. But the project leader decided to continue designing in Phase Two before completing implementation of Phase One. This was a mistake, and led to two effects. First, because the centre of action for CLT had shifted elsewhere, there was not enough person-time available to shift the Phase One implementation process along quickly, and this meant that design-team members from Phase One were left high and dry, waiting for a visible outcome from their enthusiastic work. And second, other circumstances created a large-scale change in the managerial context; the director was replaced and the locus of managerial control in the organization was completely lost for a time. Strategic direction for the project thus became even less possible, and one outcome was that staff shortages in CLT were not dealt with for far too long. It is quite possible that a more conservative style of project management would have avoided the worst effects of this situation, by focusing effort on a narrower front. With 'politics in command', the very commitment which drove the project and put gender on the agenda in the first place also jeopardized the project through a poor balance of politics and managerial pragmatics.

Women's Empowerment and the Male Status Quo

In hindsight we can say that the project leader was too close to the particular set of users involved in the Phase One design teams. These women were among the most exploited, poorly valued staff in the organization, and the CLT head was committed to delivering benefits for them at an early stage in the project. The department involved was very conservative, however, and managerial inertia bogged down the implementation, and front-line staff began to lose their sense of making progress. At the same time, to the extent that empowerment was happening for low-graded female front-line staff, higher-level professionals and managers were feeling excluded from the action. Thus, when predictable industrial-relations issues arose (health and safety of VDUs and keyboards), both the male senior manager and the male professional trade union representative seized on them in order to reassert their position. Assurances from the female staff, that industrial relations issues could easily be ironed out, were rejected by the men in favour of an interpretation which maintained the managerial/professional/gender *status quo*. The health-and-safety dispute was used to prevent staff from learning to use the system and thus practically validating both their design knowledge and their roles as designers.

Three 'Missions' in Human-Centred Systems Development

The Information Systems Use (ISU) design model was developed for a different local authority, as part of a training and staff development strategy (Hales, 1991a). The implementation plan for this model led to three 'missions' being identified, which are relevant to an analysis of the SSD case:

—developing *managers' competences and confidence*, leading to improved strategic capability and an 'opening' rather than a 'closing' management style;
—developing *'information worker' staff, jobs and careers*; and
—developing *participatory methods* for computer systems design.

In terms of the ISU model and this implementation structure, various elements are missing or underdeveloped in the SSD case.

The participatory design mission was extremely successful, producing a robust, flexible and comprehensive design within time and budget limits, but both the information-worker and management development missions went largely unrecognized. Absence of a focus on job design and career development means that the basis of participation and the role of design teams are now becoming problematic in the implementation stage. Absence of a programme of systematic management development means that it is unclear whether these job-level issues will be effectively addressed, in an organization where traditionally such 'hot' issues are not well handled.

If both these missions remain weak, the gender dimensions of the project will remain underdeveloped. Together, the three missions are intended to provide a combination of top-down and bottom-up movements and dynamics. The SSD project, being essentially middle-outwards, can be said to have succeeded in developing the bottom-up dynamic as far as is possible without better top-down support across a broad range of organizational issues and managerial domains.

The 'Management Agenda'

The CLT project leader was aware that broader organizational developments were required in order that participatory values might be thoroughly developed in the design and implementation process, and if a good (well-situated, appropriate) design and a good (effective, creative, democratized) use practice were to be the outcomes. The supplier's strategy study explicitly confirmed this understanding, presenting SSD with both a system-development agenda and a management-development agenda. The study was good, but was it good enough? Was the translated and condensed version (Hales, O'Hara and Smith, 1988) good enough? Given that whole dimensions of a human-centred approach have dropped out of sight in SSD, clearly not. Human-centred designers must also be designers of social and political 'design' practices and, while there was a recognition of this among key actors in CLT, their experience and their power to change the organization (limited by their modest middle management positions) were not sufficient to make the management agenda stick, when it came to practices on a larger organizational scale.

Theory and Managerial Practice

Are there well known, tried-and-tested organizational development methods and techniques that should have been used and which, by addressing the management agenda, would have made a clear difference to the gender outcome at the present stage? Was CLT guilty of ignorance in organization development matters? Or is it all down to power and opportunity, local contingencies and personal history, the collective competence and inertia of the whole organization? If middle managers had been better methodologists, or better versed in the human sciences, would women in SSD be certain of gaining more from the project?

The gender interest was mainly lodged in job design, role change and staff development issues, and thus (at a higher managerial level) in the policy-into-practice issues, which SSD managers are poor at handling. Certainly, if action on the items of the management agenda had been effective, the gender interest would not have been submerged in the way that it has. However, we feel that personal courage, rather than managerial technique, was the missing factor. Gender-aware organizational practice, which challenges conventional managerial practice, needs good managers who can take responsibility. We are not convinced that the social and human sciences have a lot to offer here; we are more inclined to look to the (relatively rare) skills of certain kinds of facilitator-professionals which, while theorized to some extent, are essentially apprenticeship-based, *lived* action frameworks (Egan, 1988a, 1988b, 1990).

Equal-Opportunity Practice — Plans and Situated Actions

We can identify four sites in an organization that might be seen as potential loci for gender-aware IT-related practice:

- —the formal equal opportunity and human resources policy of the employing organization;
- —general management practices, culture, style;
- —specific IT systems development techniques and methods;
- —personal commitments of key actors.

The four sites can be related to each other along two dimensions; see Table 9.1.

The rows of Table 9.1 represent *hierarchical levels* within the organization. The columns represent different *degrees of generality in knowledge*. The distinction between the two columns is closely related to that developed by Suchman (1987) between plans and situated actions. Principles/plans are more abstract, formal and explicit, and hence, publishable. The knowledges in the practices/situated actions column are more contextual and implicit, which is why we have distinguished them as 'skills'. However, in our view *all* knowledges are contextual. Convention simply ignores the 'local' and culturally-contingent nature of those knowledges that are tied to practices such as academic research; we regard them as being more 'objective' not because they are context-free, but because they belong to a peculiar and valued sort of practice, which seems to be 'universal' because it is materially abstract.

In the SSD case, the 'Principles' column was weaker than 'Practices'. The project situation throughout was underdetermined by the structures on the left, and thus those on the right were finally determining. In particular, the project-level practical structures [4] were crucial in giving the project its shape and fixing its current outcomes. Corporate-level practical structures [2] were generally too weak, relative to those within the project.

Table 9.1: Possible sites of gender-aware IT-related interests

	Principles plans and policies; 'knowledges'; official structures; formal functions; received wisdom	Practices situated actions; 'skills'; know-how; personal visions; identities; action knowledge
Corporate Level	[1] • policies and strategy statements on human resources, equal opportunities, information technology • business plans, budgets	[2] • management practices and styles [e.g. accountability, enabling and empowering] • change 'missions'; task-oriented resource groups
Project Team Level	[3] • design and implementation techniques and methods [structured design methods, participatory forms of organization, prototyping] • training and facilitative techniques and methods • job design methods and principles • equal opps strategy, in local operational terms	[4] • commitments and visions of key actors, especially top-down/bottom-up dynamics • skills and know-how of key actors • risk taking by key actors • confidence, 'personal power', assertiveness of key actors

Corporate-level structures have varied in their importance. Initially (in breaking free of the in-house contractor) they were central. But now — despite the fact that the strategy study underlined their continuing importance — they are much weaker than project level structures. Corporate-level policies and strategies have an unrealistic, detached quality seen from the standpoint of day-to-day operations. That is, corporate level 'principles' such as policies and budgets [1] have little real connection with team-level 'principles' such as methods, models and techniques [3]. Perhaps this is because the two are connected via too weak a connection with corporate-level practices [2], which in turn are weakly connected with team-level practices [4].

Because 'principles' at team level were to a large extent developed, assembled or discovered by the team in real project time, this means that even if there had been better integration between cells [1], [2] and [4], there would nevertheless have been a time lag causing corporate 'principles' to inevitably lag behind 'principles' in the front line. There is some kind of absolute difficulty here, to do with the failure of [2], and hence [1], to reflect [4]; also a problem of 'uneven development' because of lags as local learning works its way up and down the various levels or layers of organizational practice.

If the grid in Table 9.1 is seen as a window with four panes, the cleanest (most transparent) pane throughout the SSD project has been [4]; the dynamics and shape of the project have pivoted on key actors in the team. During the project, [3] has been 'cleaned', although there are still murky areas and smears.

At the present stage, it is vital that [2] should be cleaned next, or the dynamics, the embodied values, of [1] will be dissipated as 'the project' becomes 'the system' — a mere apparatus of computing power and software, distributed through many geographical locations and integrated into practice in various *ad hoc* ways. It can be said that [2] is an area of crucial weakness for the organization as a whole.

We suspect that a similar story would emerge in most large organizations. Because of poor organizational learning and weak 'bottom-up' processes, [1] will have a weak connection with [2]; for the same reasons, and also because of inadequate management practices at a personal level, [2] is likely to have an unsystematic connection with [4]; and the specialized, complex mix of knowledges represented by [3] is extremely underdeveloped, at the general level of 'the literature' and professional training (on both the contractor side and the client side of the IT development relationship). Thus human-centredness and gender awareness in an IT project will in most situations pivot on [4] — the commitments, vision and personal power of the client-side project team; and they will be limited to precisely the same extent as the power of a middle-level staff group within a complex bureaucratic organization is limited.

Conclusions: Politics, Design and Management

Can We Call This a Human-Centred Design Project?

'Human-centred' (HC) can have many meanings. Our understanding from others, particularly in Britain, is its combined emphasis on personal identity, organizational settings, managerial dimensions and methodological issues (Hales, 1991b). Our interpretation is that:

- HC embraces design, implementation and evaluation, as a *continuing organizational process* within a concrete network of users' and producers' practices;
- HC centres on designing a system of local design practice. That is, it is concerned with *organizational and methodological issues* such as the make-up and methods of design circles;
- local design practices, in turn, design *use practices as distinct from mere technological artefact-systems*. That is, HC design designs jobs and roles too;
- to produce adequate knowledge of a complex object, it is necessary to produce a complex Investigative Subject of a kind that probably does not exist within a given bureaucratic, fragmented organization. That is, *political 'organizing'* is part of HC design;
- while all good design is covered by the above principles, 'HC' has historically signified: a) a striving for the cognitive and emotional *empowerment* and significant *recognition* of final users as designers (this is the 'participatory' mission), together with all their relevant knowledge and know-how; and b) a recognition that systematic inequality — especially in class and gender terms — is an aspect of all practices involving technologies and work, and that this places definite limits on what can be accomplished through *professionalized* design activity;

— HC design accepts that *living labour* (human action) — as distinct from machine systems or formal routines — is the source of 'service', 'quality', 'value added', and all such characteristics of economic activity. Thus, *all* the forms of knowledge that frame people's action in work are relevant in design, including tacit knowledges, personal identities, work roles, informal knowledges and skills.

The SSD project has been good on 'practice rather than technology' and, although hampered by the general weakness of relevant theoretical understandings, has succeeded quite well in developing local knowledge through the combined methodological efforts of the project team and bought-in consultants/facilitators. However, because of the demands of keeping a large IT project on the rails, and because human resource development is a weak focus in the organization's management culture, 'design of jobs and careers' has been a very understated theme. Project management — especially the design of the design teams — has been strong on design as organizing, empowerment of final end users within the design process, tacit and informal knowledge and politicized management, but effects in the political dimension have been circumscribed by the middle-level status and limited personal resources of the main actors.

So, can we call the SSD design project 'human centred'? Our view is that human-centred design cannot be guaranteed by any methodology or design model for two reasons. First, there can be no such thing as a human-centred artefact — only a human-centred *practice*. Thus human centredness lies in the continuing struggle to realize the values of the above definition, rather than in any given engineered system or organizational outcome. And second — a rider to this — a human-centred technology-use practice only remains so if it is part of a continual (human-centred) design process, being periodically recycled through explicit review and redesign phases.

In these terms, the SSD case can certainly be called a human-centred design project. Given its incompleteness and some of the weaknesses identified in this chapter, however, we are unable to say whether the practices surrounding the IT system will remain human-centred over the coming years of use. In particular, we cannot say for certain whether the currently implicit gender dimensions of human-centred use, redesign and job design will emerge strongly.

The Empowerment of Information Workers

An essential dimension of human-centred IT design is the (re)design of jobs and careers for 'information' workers, most of whom are women. As yet, however, in SSD, managers' agenda does not include using the technology, as it emerges on the ground, to create openings for staff development, for learning and for structural change in work practices. In CLT's vision, technology was supposed to be 'enabling', which meant enabling people to participate at a level that was meaningful to them — for example, improving their personal skills. This aspect has been lost among problems of designing a complex IT system (an apparatus) and managing a big systems-delivery project. Certain staff were enabled as participants in the front-end design process; within this process they were empowered as actual contributors to new kinds of activity. The system that resulted is, in turn, capable of enabling a whole range of new activities in SSD, but the process of empowerment of the whole of SSD — logically implied by the way the system was designed — is a project to which top management have yet to commit themselves.

Can staff see the opportunities now? Do they believe in them sufficiently to make a fuss about them? Do they believe that their bosses and the professionals can deliver a real change in power structures within the office environment? Are staff willing to take the risk of becoming involved as designers/changers (and thus responsible, to some degree for future failures), or will they just remain employees and blame 'the management'? These questions identify the ground onto which the project will have to move, if the gender-aware commitment is to be met.

Academic Research and Learning by Doing

This project can be read as a failure of academic knowledge and progressive ('participatory') ideology; neither were able to take the project where it wanted to go. There is no reason, however, why progressive-minded academics should be embarrassed or disheartened by this. We feel that human-centred design, because it is about politics as well as technique, is inevitably a vocation rather than a profession (certainly not a bundle of design skills) and that it is therefore impossible to legislate for or pre-specify human centredness; no abstract structure of methods, techniques, models and policies (no 'principles', no 'plans') can guarantee it. Human centredness lies in the live structure of acts, behaviours, practices. The same can be said of gender-aware design, seen as a form of HC practice informed by a particular interpretation of politics.

Thus, because academic work — especially in science-related areas — is so heavily dominated by values of detachment, abstractness, pre-specification and formalization, so far from the value of self-reflexive practical action, there is a deep difficulty in making academic knowledges serve real-world practice. Generally, they are neither the right shape, nor in the right place. Some of the practitioners in the SSD project were committed to theorized reflection on practice and had some active connections with current theoretical work. But there is a practical constraint on how much knowledge from elsewhere can be shipped into a given situation. If the overriding commitment is to get some things done — to make and seize some opportunities for change — then wheels will have to be reinvented and 'obvious' mistakes, will have to be made.

Nevertheless, if [3] in Table 9.1 — the 'professional', researchable, publishable apparatus of design, project management and facilitation of organizational change — were stronger, it could better support the situated action of people in projects [4]. We do think that it is important for systematic research to be done on the whole range of project-level 'principles', but academic researchers should resist the temptation to claim that their generalized knowledges are, or even can be, 'human-centred'. This is a term to use very, very carefully, and only of *practices* of actual social/organizational change, as distinct from abstracted knowledges.

Academics are currently doing important and useful work, resulting in publications that carry practical insights: for example, into collaborative design techniques (Greenbaum and Kyng, 1991), or the relevance of 'web' and 'activity' models of the design Object (Kling, 1987; Kuutti, 1991), or design-in-use in office work (Clement, 1991). Eventually it may be that the graduates of revised education programmes will populate the systems design professions and bring to bear a more helpful understanding of the social and political nature of design practice and its objects. In the meantime, user-side actors will have to manage in as reflective and theorized a way as they can.

Ideology and Human-Centred Design as a Vocation

We are disturbed by a final conclusion which can be read into Table 9.1. If [2] were stronger — corporate-level management practice — it could effectively support both [1] (the policy/planning process: important because of its 'publishability') and (by actively structuring 'space' for it) [4] — competence in local, project-level actors. Recognizing this, we must be prepared to believe that despite the greater likelihood of public sector organizations having an explicit formal commitment to equal opportunities, more headway might be made with gender-aware IT development in a commercial organization with very capable managers.

Managers in SSD were unable to *deliver* either cross-departmental strategic coordination of action, or a process of empowering development that reached wider than the small transient population of design teams. It seems to us that under other banners ('total quality', for example) some private sector organizations may actually be doing more to empower their staff than most 'right-on' local authorities do. It is easily possible that, let's say, a City finance house, with a market-driven 'customer facing' strategy and confident managers with facilitative skills, could be producing more real change — more 'action space' and personal control for female front-line staff — than will happen in SSD for years yet. Can this be a bad thing for those women?

For socialist-feminist, public-sector workers, this is a tough conclusion to swallow, and for those with a vocation for HC, it makes the question of 'where's the best ground to work on?' that much more difficult. Experienced consultants and self-starting professionals know that very often there is not much point in working for an incompetent client or boss when you could be working for somebody more capable; and it is beginning to be apparent that the ideological conditions for successful HC development are more complicated and contradictory than we have tended to recognize.

References

BOLAND, R. and HIRSCHHEIM, R. (Eds) (1987) *Critical Issues in Information Systems Research*, New York, Wiley.

CARROLL, G. *et al.* (1989) *A Method for Developing Standards of Competence in the Constructive Use of IT*, final report by Mainframe to the IT Industry Lead Body (End User Group), Training Agency, Moorfoot, Sheffield, England.

CLEMENT, A. (1991) 'Designing without designers — More hidden skill in office computerization?' in ERIKSSON, I.V., KITCHENHAM, B.A. and TIJDENS, K.G. (Eds) *Women, Work and Computerization*, Amsterdam, Elsevier.

EGAN, G. (1988a) *Change-Agent Skills A — Assessing and Designing Excellence*, San Diego, University Associates Inc.

EGAN, G. (1988b) *Change-Agent Skills B — Managing Innovation and Change*, San Diego, University Associates Inc.

EGAN, G. (1990) *The Skilled Helper — A Systematic Approach to Effective Helping*, 4th edn, Pacific Grove, California, Brooks-Cole.
EHN, P. (1989) *Work Oriented Design of Computer Artifacts*, Hillsdale, NJ, Erlbaum.
ERIKSSON, I.V., KITCHENHAM, B.A. and TIJDENS, K.G. (Eds) (1991) *Women, Work and Computerization*, Amsterdam, Elsevier.
GREENBAUM, J. and KYNG, M. (Eds) (1991) *Design at Work — Cooperative Design of Computer Systems*, Hillsdale, NJ, Lawrence Erlbaum Associates.
KLING, R. (1987) 'Defining the boundaries of computing across complex organizations' in BOLAND, R. and HIRSCHHEIM, R. (Eds) *Critical Issues in Information Systems Research*, New York, Wiley, pp. 307–62.
KUUTTI, K. (1991) 'Activity theory and its applications to information systems research and development' in NISSEN, H.E., KLEIN, H.K. and HIRSCHHEIM, R. (Eds) *Information Systems Research — Contemporary Approaches and Emergent Traditions*, Amsterdam, Elsevier, pp. 529-49.
HALES, M. (1991a) 'A human resource approach to information systems development — The ISU (Information Systems Use) design model', *Journal of Information Technology*, **6**, December, pp. 140–61.
HALES, M. (1991b) 'User participation in design — What it can deliver, what it can't, and what this means for management', paper for SPRU/CICT workshop on *Policy Issues in Software and Systems Development*, Brighton, July.
HALES, M., O'HARA, P. and SMITH, G. (1988) *Progress by Design — An IT Strategy for the Social Services*, Camden Social Services Department.
ILLICH, I. (1973) *Tools for Conviviality*, London, Calder and Boyars.
NISSEN, H.E., KLEIN, H.K. and HIRSCHHEIM, R. (Eds) (1991) *Information Systems Research — Contemporary Approaches and Emergent Traditions*, Amsterdam, Elsevier.
SUCHMAN, L. (1987) *Plans and Situated Actions — The Problem of Human Machine Communication*, Cambridge, Cambridge University Press.
VEHVILAINEN, M. (1986) 'A study circle approach as a method for women to develop their work and computer systems', paper for IFIP Conference, *Women, Work and Computerisation*, Dublin, August.

Chapter 10

The Universes of Discourse for Education and Action/Research

Fiorella de Cindio and Carla Simone

Abstract

In this chapter we propose the Universes of Discourse as a tool for analyzing the various dimensions of work and its computerization, and describe their application for developing a framework for identifying and discussing the specific work attitudes and practices of women. Finally, we illustrate how this framework has been used in various teaching experiences and in action/research programs devoted to women.

Introduction

Some years ago, during research devoted to the development of computer supports for office/clerical work and the assessment of the impact of these supports, we developed (in cooperation with a colleague) an analytic tool called 'the Universes of Discourse'. This tool was designed to interface with the different perspectives from which both designers and users look at and describe the Office Information Systems in which they are involved. These perspectives can be recognized by considering the different lexicons users and designers utilize in describing office phenomenologies (De Cindio, De Michelis and Simone, 1987; Simone, 1987).

Meanwhile we had started to reconsider our professional competence as teachers and information technology researchers from a gender perspective (see, among others, De Cindio and Simone, 1985; Catasta, De Cindio and Simone, 1986). Once we had realized the effectiveness of 'the Universes of Discourse' for educational and work (re)design purposes, it was quite natural to experiment with them in education and action/research activities devoted to women in various settings. Principally these settings have included training courses for working women, organized on behalf of unions and companies, and the development and analysis of organizations within research directed at affirmative action[1] programs for women in the workforce (De Cindio, 1986; De Cindio and Simone, 1985, 1988; Catasta, De Cindio and Simone, 1986; Osimo, 1988; Osimo and Simone, 1989). This led to the development of a framework for understanding women, work and its computerization, broadly presented in Tijdens, Jennings, Wagner and Weggelaar (1988).

In this chapter, the first section explains and illustrates the Universes of Discourse. The second section reports on their use in training and in the analysis and (re)design of the organizational settings involved in affirmative action programs, and in addition, discusses the impact on both those who use them and on those who reflect upon them.

A Confusing Terminology

Both designers and users involved in the design of socio-technical systems know that they and others frequently describe the same phenomenon using different linguistic dimensions or 'lexicons'. For example, posing the question 'What are you doing?' to a person using a VDU may prompt the following responses:

I am typing on a keyboard and looking at the screen, and my eyes ache.

I am using a work-processor which is not provided with an 'undo' command, thus some mistakes become disasters!

I am drawing up the monthly budget report. I have to finish it before tomorrow's committee meeting and I am running late because some of the data is wrong!

Mr Jones asked me to prepare the budget report. I probably need Mrs Kleene's help to describe the marketing aspects too.

I'm angry as I agreed to do the report for a colleague of mine. It will never happen again!

Each of these responses is meaningful *per se* but contains quite different information about the phenomenon under consideration. The first one speaks of physical objects and physical movements; the second one a facility and its use; the third one refers both to specific steps (the report preparation and meeting) within a more complex process (budget handling), and to their mutual relationships (time constraints). The fourth emphasizes the motivation (a request by Mr Jones), and future cooperation with others (Mrs Kleene); while the last response highlights distress over having been involved in an unpleasant job.

The words, or 'lexicon', used in each answer stress different dimensions of the work, different perspectives from which the phenomenon 'using a VDU device' can be considered. In reality, people are more likely to mix the various linguistic dimensions, since they live in all of them, holding the five perspectives concurrently. More probably, the answer obtained would be something like: 'I'm typing on the keyboard, but I'm very late in finishing the work I promised for tomorrow'.

While this mixing of dimensions, and of lexicons, may be natural, it can nevertheless lead to difficulties when the language is being used to communicate or express needs, or to define problems and their solutions, rather than just to chat. For instance, users often ascribe their problems and uneasiness with factors such as eye fatigue and general stress after a working day. This can lead system designers to seek solutions along the wrong dimension, that is, to assume that tiredness and stress problems would be solved by changing the VDU device, and/or by reducing the time spent using it. Instead empirical research (Novara, Rozzi and Sarcinelli, 1983) shows that in a number of cases, only the redesign of a set of procedures can actually solve them. In general, a poor formulation of the problem may also lead the designers to fail to address the actual need for an overall redesign of the work, mistakenly considering only one dimension of a problem while disregarding the possible consequences on the other dimensions. For example, a group of researchers may suggest that: 'To avoid spending hours trying to use the copier on the first floor, we would like have one on our office floor'. While the 'natural' solution of buying a new copier and locating it on the researchers' office floor addresses the technical or logistic aspects, it nevertheless disregards the organizational ones, such as: who is responsible for maintenance of copiers located outside the Secretariat?

What is missing is a reference system in which the different aspects of problems and solutions can be compared on a sound basis, allowing the comparison of commensurate values with the aim of finding a good balance between all the dimensions. What we propose (the idea was first developed in De Cindio, De Michelis and Simone, 1987), is that the five 'lexicons' allow one to recognize five linguistic dimensions for analyzing (office) work, skills and technology.

The Five Universes of Discourse

The five responses given as examples above, are constructed of different types of words, different 'lexicons', which characterize different Universes of Discourse. Each of the responses describes the phenomenon considered by using a different set of terms and of relationships binding them.

The five universes definition

The first of the universes is the *Physical Universe* where the response: 'I am typing on a keyboard and looking at the screen, and my eyes ache' makes sense. The lexicon used here describes the phenomenon in terms of both the physical environment in which the work is done (for example, the kind of furniture, the layout of facilities, etc.) and of the movements and sensations of the people performing the work.

The second is the *Operational Universe*, where the response: 'I am using a word-processor which is not provided with an "undo" command, thus some mistakes become disasters!' makes sense. The lexicon used here describes the phenomenon in terms of the kinds of operations and facilities that can be used to perform these operations, that is, writing a text using a word-processor. The type and size of the operations used depends strictly on what is considered a significant unit of resource/facility.

The third universe is the *Procedural Universe*, where the response: 'I am drawing up the monthly budget report. I have to finish it before tomorrow's committee meeting and I am running late because some of the data is wrong!' makes sense. The lexicon characterizes the phenomenon in terms of products (reports, letters, invoices, orders, budgets, etc.) and of the input/output transformations performed to produce these products.

The fourth universe is the *Commitment Universe*, where the response: 'Mr Jones asked me to prepare the budget report. I probably need Mrs Kleene's help to describe the marketing aspects too', makes sense. Here the lexicon characterizes the phenomenon in terms of both the conversations necessary to define and redefine people's mutual commitments, and of the roles which characterize the group structure, as defined by the conversations in which they are involved.

The fifth and last universe is the *Individual Universe*, where the answer: 'I'm angry as I agreed to do the report for a colleague of mine. It will never happen again' makes sense. The lexicon in this universe characterizes phenomena in terms of people's perceptions, attitudes and expectations.

While the five Universes can be used to describe any kind of activity within organizations such as offices, families, teams, etc., our focus here is on the office work environment, which we will discuss from a woman's perspective.

Let us point out here that it is the Procedural and Commitment universes which provide the linguistic framework for characterizing group work, and also for characterizing each specific office in terms of the rules governing both the execution of its tasks and actions, and the definition and fulfilment of its commitments.

The phenomena occurring in the office can be described and evaluated in each universe separately, and there is no dependency or hierarchy between any two of the universes. At a first level of abstraction, using a metaphor derived from geometry, the five Universes define a five-dimensional space in which each coordinate autonomously contributes, in a complementary way, to the qualification of any phenomenon in the space. We will use 'dimension' as a synonym of 'Universe' in the following discussion.

Thus, the five Universes can be used as a grid, or structured framework, in which the parameters for comparing problems and solutions, in both a qualitative and quantitative way, can be identified. This grid can be used to study:

- the *supports*: resources, facilities and technology;
- people's *skills*; and
- the *problems* and *disciplines* necessary for studying them, in addition to the parameters for comparing problems and solutions in both a qualitative and quantitative way;

which make sense in each universe.

With regard to technology, we will refer mainly to computer-based (CB) technologies, and in particular to those supporting clerical work.

The grid based on the universes

In the *Physical Universe*, the supports to consider are the office equipment (desks, cabinets, partition panels, lamps, furniture etc.) and their physical interfaces; which for IT technologies consist of mice, touch and high resolution screens, vocal analyzers etc. Problems concerning the physical environment (office layouts, illumination, etc.) can include the *consequences* of using the facilities, such as physical fatigue, distress due to bad postures, eyeaches, etc. The significant workers' skills in this dimension include: their resistance to physical fatigue, manual precision and dexterity. The disciplines involved are: work environment architecture and ergonomy, while the assessment in this universe is based on parameters such as the mutual dislocation of objects, materials and machines, their 'ready-at-handness', and the illumination, ventilation and sound-proofing of the environment.

In the *Operational Universe*, the technology includes the applications typical of personal computing (word-processors, spreadsheets, databases, graphics, personal agenda and project management packages etc.), applications related to specific professions (accounting, budgeting, CAD/CAM, scientific applications, Decision Support Systems etc.), query languages and fourth generation languages; and finally, the interfaces with the applications typical to the Procedural and Commitment Universes.

Parameters, significant in the Operational dimension concern both the facilities available and the operations which these facilities support. The operations parameters are their relevance, frequency and duration, while the facilities parameters are: how easy to learn and to use they are; their access/response times; the richness, scope, integration and flexibility of their functionalities; and the availability of built-in functionalities supporting error/exception handling. As proposed by Novara, Rozzi and Sarcinelli (1983) and Bagnara (1987), we call the generic problem which can arise in this dimension *stress*, which, from time to time, can be characterized in terms of the parameters listed above. Thus we can observe stress due to the excessive duration of an operation, to the use of a word-processor without an undo command, to the use of an application with very slow response times, and so on. The discipline useful in dealing with these problems is cognitive ergonomy. Examples of the workers' skills significant in this dimension are: competence, experience, level of autonomy in the use of the various facilities, the number of applications of similar and different complexities with which one is familiar, and resistance to stress.

The support technology significant in the *Procedural Universe* includes:

— all the classical EDP applications (for stock control, personnel, billing, customers and suppliers),
— possibly integrated in the Management Information Systems and with data banks (medical, statistic, territorial data banks etc.),
— The forms/documents handling applications,
— The integrated CAD/CAM/CIM applications, etc.

In this universe the assessment parameters refer to the qualitative and quantitative aspects which have to be balanced to achieve efficiency and effectiveness. They concern:

— The richness, scope, integration, flexibility and complexity of procedures: e.g. the types and numbers of facilities used, and of products and roles involved;
— Their control structure: the degree of sharing of and competition over resources, synchronization of activities by mutual exclusion, blocking or concurrency and resources access times;
— Their flexibility: the availability of features supporting error/exception handling and the effects of unforeseen events;
— The level of autonomy in the solution of conflicts: for example if the criteria are predefined or partially discretionary, or if the conflicts solution is a local or a cooperative process;

— The nature of the communication involved: synchronous/asynchronous, active or passive, structured or unstructured;
— Their duration and costs;
— The timeliness, immediate usability and form adequateness of the input products from other procedures, making them easily usable or not, in the considered segment of productive process.

In this Procedural Universe, the problems are due to the difficulties in finding the best balance between the values of different assessment parameters which can be contradictory. For instance, the highest concurrency does not give the best results, since it implies high complexity and hence a high probability of error. The discipline for analyzing work in the procedural dimension is scientific work analysis as historically developed and applied both to productive and to clerical (Simon, 1977) work. The most precious skill in the procedural dimension is the ability to control the flow of activities and variances, for example, the ability to isolate the current problem so that the overall procedure can restart, or the ability to find the best scheduling and resource allocation.

The most frequently used technology in the *Commitment Universe* has been, up to now, the telephone with all its extensions (telex, telefax, teletext, etc.). In some environments (highly developed countries and large multinational companies) we also observe the broad use of Electronic Mail Systems based on Geographic Telecommunication Networks and on Local Area Networks. Further, in recent years people have given increasing attention and effort to the development of software to support the communication, conversation and commitment networks of which all non-routine work consists. This software supports the work group dimension (often referred to as 'groupware') or in other words, people's cooperation. The application spectrum is really wide: from applications to handle asynchronous communication among users (to start, close continue, store, retrieve, filter communication on the basis of their interests), to applications supporting synchronous communication possibly in distributed environments, for example, (tele)-conferencing, (distributed) meetings, co-authoring systems; from packages supporting the management of a group of people involved in a project (project management tools) and the coordinated decision processes (group decision support systems), to applications for combining various communication media in an integrated framework, exploiting different man-machine interfaces (including restricted forms of natural language); and finally tools combining various of these possibilities. For a comprehensive review of these innovative proposals, see Greif, 1986; Greif & Suchman, 1988; Bikson & Halasz, 1990; Olson, 1987; Greif & Ellis, 1987; and Ellis, Gibbs and Rein, 1991.

In this Commitment Universe problems concern the efficiency and effectiveness of negotiation, which in turn is related to the complexity of the office conversation network, and to the definition of the office roles. The complexity of the office conversation network depends on the difficulty of identifying and reaching the partner, on the adequacy of the support technology used, and on the number and frequency of conversations and their subconversations. Role definition is crucial, as it determines the level of flexibility of the overall office and the autonomy of the roles themselves, in terms of the domain of possibilities each one has. This role definition determines the degree of visibility of the overall process being governed, and the nature of such government, be it coordination, control, support or consultancy. These issues determine the degree of ambiguity and conflict between roles. All the above mentioned 'groupware' technologies aim at supporting their users in dealing with these problems, while the existing technologies (telephone and EMS) leave all the coordination to the user.

Some of the parameters significant in this dimension have been cited in the above description of the problems; a more complete list can be found in Simone (1987). What characterizes workers' skills in this dimension is their autonomy, and their ability both to live in complex human relationship networks, and to interpret their role with competence and flexibility. In this universe the main discipline is phenomenological work analysis, of which Winograd and Flores' study, (1986) represents the most complete attempt to develop this discipline for the analysis of (office) work.

In the *Individual Universe* it is meaningless to speak about support technologies. The problems here are related to people's personal reactions to the organizational, technological and social settings which constitute the work environment, as it results from the composition of the factors occurring in the four previous dimensions. The related disciplines are the sociology of work and psychology, subjects outside the authors' professional competence. It has always been important for us to cooperate with researchers working in these disciplines, in order to extend this Universe to show the same level of analytical detail as the others. Nevertheless, our efforts in this direction were not completely satisfactory. In fact, while the 'in the field' cooperation with some individuals was profitable and stimulating, it did not produce a reusable and systematic understanding, such as a set of parameters or analysis methods, as happened for the other dimensions.

Using a metaphor derived from geometry (see Figure 10.1), the five universes define a multi-dimensional space, in which each coordinate autonomously contributes, in a complementary way, to the qualification of the phenomena in this space. We have used this framework in a number of different contexts for:

- Educational purposes, in order to make people — managers and employees, professionals and trade unionists, men and women (see the next section) — aware of the combination of activities of which their work consists.
- Evaluating the different dimensions of (office) support technology usability (De Cindio, De Michelis and Simone, 1987).
- Characterizing a set of quantitative and qualitative parameters to be used in office analysis (Simone, 1987).
- Defining new ways of acquiring linguistic competence by young children and, therefore, as a different characterization of linguistic knowledge (De Michelis, 1987).

Others (Butera, 1987) have used this framework for a characterization of professions. Having been utilized in many ways by many people, the Universes of Discourse tool has proven itself to be both powerful and useful. In particular, the tool has proven its worth in the analysis of office work and its computerization, where its use avoids the misleading mixing of different dimensions and this avoids the problematic situations described in the first section. In addition, as we shall see in the following section, this tool constitutes a framework for reflecting upon some of the specific characteristics of women's work, and the changes which occur as a consequence of computerization.

Figure 10.1: The five-dimensional space

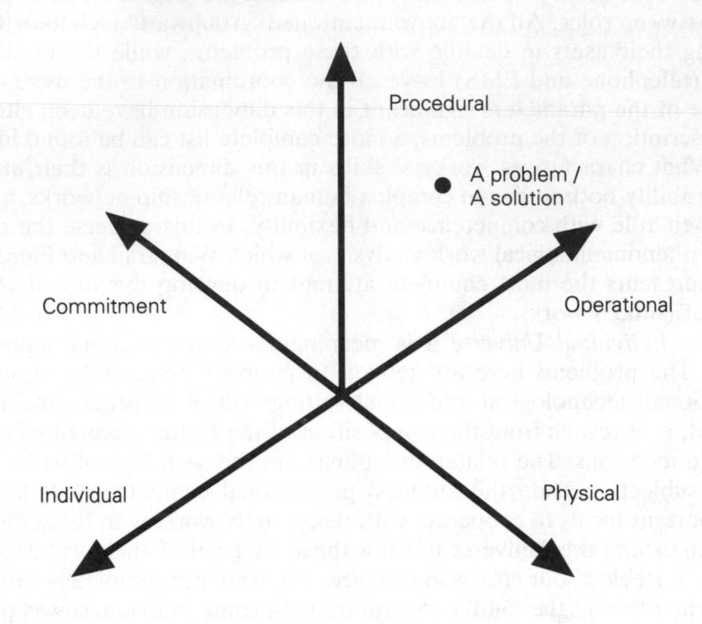

Teaching the Universes

Since we developed the Universes of Discourse schema, we have used it in many courses and conferences with which we have been involved. These courses and conferences have been of widely differing types, ranging from refresher courses on office automation organized by unions, to specialist courses about 'groupware' for company employees. Further, the audiences addressed have also varied widely, at times being composed entirely of women. Examples of the above include: courses within Affirmative Action programs; meetings on the impact of new information technologies; and within conferences aimed at developing gender-oriented perspectives on science and technology.

In each case, where the Universes of Discourse was used with the aim of providing a reference frame and a precise terminology, the results were extremely satisfying, in that participants recognized it as a means by which both to distinguish and to compare phenomena which had previously remained confused and tangled. In particular, audiences of women appreciated their ability to use the tool to deal with both the attitudes of, and towards women. We return to this theme below, after dealing with how the grid of the Universes is introduced and discussed. For the sake of a simplified account we shall assume that the presentation discussed is directed at a group of workers employed in the same office of a large organization, who have at least four half-day sessions available for the course (although positive results have been obtained in far less time). The type of organization (more or less bureaucratic) and the productive field (services versus goods) in which the organization operates, do not affect our teaching methodology; although they can influence the participant emphasis on specific work dimensions (for example, the procedural universe vs the commitment universe). One of our basic teaching principles is to contrast any stereotype and to facilitate the reflection on all dimensions in any type of organizational setting.

Usually we adopt a *three phased approach*:
First, we start by eliciting from the audience a description of the work situation, of the problems encountered, and of the equipment used. This phase has two functions: first, it supplies the examples to illustrate the use of the Universes in later phases, and second it encourages the group's active participation in the process and helps them to recognize us as group 'secretaries'. The secretary role is to coordinate the discussion, ensuring the participation of all, and to record everything that is said in such a way as to allow, (after some 'cut-and-paste'), a collection of statements such as:

> The terminals are put on desks and not on suitable tables; they are too high and cause backaches.
>
> Before I would type for half a day or more. Now I have more time in which to help others.
>
> Those receiving goods may not let us know for some days which can delay payments for up to a month.
>
> When an invoice doesn't tally, a telephone call would suffice to fix the problem.
>
> I like things the way they are. I don't have any suggestions about how my work could be improved.
>
> (Note that these statements are recorded verbatim during real situations.)

Then we then point out that each statement refers to a different dimension of the work. The first relates to physical aspects, the second to equipment and operations and so forth, thus informally introducing the five Universes. As we proceed, two other features are pointed out: first, that in each dimension we can identify what characterizes the work, the technology involved, the problems presented and the skills required; second, that this placing of things into a specific dimension is the result of working on the 'rough' statements which mix different dimensions together. At this point the group is ready for a precise presentation of the Universes, as outlined above. Here a helpful exercise is to use the Universes to 'read' the recent evolution of the technology, a difficult exercise for people with limited experience in this field. For example, the pattern which companies go through — from EDP to Personal Computing and Individual Productivity Tools, toward a computerization which also considers the communication aspects — can now be read as follows: an evolution from a computerization mainly set in the Procedural Universe, to a computerization which sets side by side with this, the applications typical of the Operative Universe and sometimes, the first applications significant in the Commitment Universe. The grid presented in the previous sections shows a maieutic capability, in the sense that the corresponding organizational changes and the consequences for workers' skills become much more visible to the audience.

The third phase of the course involves the active use of the Universes to deal with the particular problems which are the reason for the intervention. The two typical outcomes of this phase concern a) the development of a gender-oriented perspective which values rather than negates female attributes, and b) the analysis of the target organization to identify, together with the women, possible changes and reorganizations (as often required when the course is part of an affirmative action program). The first outcome is discussed in the next section, moving on to a consideration of the second in the one after that.

Attributes Specific to Female Work

In order to develop a gender-oriented perspective which values rather than negates female attributes, the women of the audience are asked to use the framework of the Universes to examine these female attributes. The grid we are going to present is the outcome of our elaboration, improved by women's comments and suggestions. However, before doing this, it is worth noting the following:

First, we are aware that some men show 'feminine' attitudes and vice versa. Furthermore, woman/man is not an abstract category, as each woman/man has her/his own identity and specificity, and her/his own 'culture of work'. Nevertheless, we choose to take the risk, since, in our opinion, this is the only way of checking whether 'difference versus homologation' is merely a slogan, or an effective strategy.

Second, the following grid is more oriented toward identifying characteristics 'in positive' rather than 'in negative'. This follows from the above-mentioned goal: if diversity has to be checked as background for an effective strategy, one must first find some advantage, and only afterwards cope with the possible disadvantages. If the first step fails, then the second is not useful. The women we met shared this approach; first they attempted to identify their points of strength, which made them more open to also recognizing their weaknesses.

Third, in presenting the grid a * denotes the abilities which seem to be directly imported into the work environment from the home/family environment. This is a first attempt to distinguish those attributes which women acquire because of gender differences in socialization, from those which are 'naturally' female. In fact it is strategic to recognize when a factor derives from the 'dual role' which women perform.

In the *Physical Universe* two characteristics specific to women's work are manual precision and dexterity. They are well-known and often used against women, who are asked to execute highly repetitive jobs, since they are perceived to have great endurance for physical stress and constant powers of concentration, especially when doing jobs where physical strength is not involved, but where other senses: sight, touch, and sometimes hearing, are heavily stressed, such as in the clothing industry, in some branches of alimentary industry and in electronic productions. In the latter industry, women are heavily exploited in the short time that they are employed, that is, until a more sophisticated technology is developed!

In contrast, the attention women pay to the physical context of the work environment* is usually not considered an important factor, and open irony is often addressed to women who put plants and other ornamental objects on top of the machines which they use, or spend time rearranging things on their desks.

Nevertheless, such attention to physical context can become a productivity factor (in the next universes) because a pleasant and tidy workplace enhances comfort and reduces the time spent finding lost objects or papers.

Finally, the attention women pay to health problems* could be used as a focus for the physical ergonomic problems related to IT technology. This topic should not exhaust the health problems related to technology, as sometimes happens, since it covers only one dimension, but on the other hand it should not be neglected, as too often happens.

We are not suggesting that enormous efforts should be made to better account for manual precision and dexterity, since these stressing characteristics will be embedded in very reliable, automatic high-precision machines. Nevertheless, we would not exclude taking advantage of women's work experience and competence for designing the introduction of these new machines. Women could act as consultants for organizing the logistics of the new work environment, for instance, for optimizing supply and for designing the kind of control to be applied to the new system (perhaps a weak and spread control instead of a dedicated attention?). This strategy could both save a number of women's jobs, and build automatic production on the basis of a year's concrete experience of work.

In the *Operational Universe* the main specific character we have identified refers to a difference in the relationships with objects and machines. For men they are mainly a status-symbol to possess: such as a car; this is also the case for computers which are 'exhibited' on the desk top. In contrast, women evaluate the objects and tools which they use in terms of the support they give. Women do not idealize computers in the same way that they do not idealize washing-machines and other domestic appliances.

This more pragmatic, concrete attitude often gives rise to a greater capability for exploiting the various tools, from possibly enriching a low-level training, to an ability to accept their features, to finding 'tricks' for dealing with even the worst ones. More generally it leads to a greater endurance of the psychological stress, which can be observed in workers forced to use software tools with a low-level of usability (De Cindio, De Michelis and Simone, 1987).

Sometimes women develop an affection for the machines which they have used for a long time; for example women working in the clothing industry may become angry when the management decides to substitute new technology for the old machines. This reaction can lead to a rejection of the new technology, but can also be viewed as a reaction against the waste* so common with IT technologies. The emphasis here and in the following discussion of this topic, draws upon the Italian situation where, as in all historically non-rich or recently enriched countries, people, and working women in particular, are accustomed to the problems presented by few resources, tools and technology.

Starting from the typical manner in which women use all the available facilities to support various office work operations, we would urge them to embrace IT technologies, so that their original skill can underpin a new profession: from typewriters to experts of wordprocessor applications, from archivists to experts of the textual spread-sheet packages, from graphic workers to computer graphic experts, and so on.

In the *Procedural Universe* the characteristics mentioned in the Operational one result in women showing greater reliability and attention to avoiding waste during the production process. Reliability plays, in general, a major role in enhancing the quality of the product, while a reduced amount of waste enhances overall productivity. We would attribute a big responsibility to women in this Universe. For a long time people were inclined to consider routine tasks as the major part of office work, with each invoice, each order equal to the previous and to the next ones, except for minor changes (the variable fields of a record). Now people are discovering that each instance or object is unique.

Women (the large majority of the employees), who have had to cope with rigid procedures and prescriptiveness, have already personally gained this experience and can say therefore which parts of the implemented procedures are effective, and which need to be changed and redesigned. They can exploit this acquired competence when a procedure redesign is undertaken to reduce the rigidity and prescriptiveness, possibly reusing part of the amount of existing software (a crucial task!).

Other characters can be viewed both as a means to overcome the rigidity imposed by the procedures and as ingredients necessary for working in the *Commitment Universe*. Since it is difficult to present them separately, we will describe some of them by discussing how each characteristic belongs to the two dimensions.

Let us begin with the ability to organize and perform a number of activities concurrently by applying a kind of weak control over them and optimizing the time spent,* a factor typically imported into the work environment from the home/family environment. Let us consider a woman returning home in the evening; she has a number of different jobs to perform in a short range of time: cooking, looking after the children, starting the washing-machine, tidying up the house. Everything should be accomplished around dinner time. She starts more or less all the various activities concurrently and applies to them a weak control, intervening when necessary (that is, when some breakdown occurs) by for instance, rescheduling her activities. We agree with the claim that: 'Women execute various household jobs in a non-hierarchical and apparently disordered way'. Isn't this apparent disorder an important way to introduce concurrency and, therefore, optimize job execution? Furthermore, in order to deal with possible crises (children's illness, two of their husband's friends at dinner etc.), women have always had different kinds of solutions/supplies available, but the minimum possible in order to avoid 'waste'.

Many women have recognized that, however unconsciously, they actually use the same strategies during their time at the office. In the Procedural Universe, if a woman is involved in more than one procedure, her first concern is to identify the 'most effective' scheduling of them by giving the highest priority to the tasks which if not completed will hold up others. In the commitment dimension this ability is crucial for living effectively inside the complex network of interrelated conversations opened to deal with break-downs (Winograd and Flores, 1986). Women who often have to handle double work loads, but want to reduce the time involved, have developed a culture which reduces time wasting since time is a precious resource. The large majority of women consider this a positive ability, which they mention as a major point of difference with respect to men's attitudes to work (they often claim that men are 'monoprocessors'!). They are also aware that such strategies allow them to be more productive than men, despite the fact that this greater productivity receives little recognition.

Another characteristic which is related to that described above, can be called the nurturing attitude.* Women derive from the role played in the home a capacity to consider it their duty to solve the problems of others. This characteristic is frequently used in an employment context, for example by managers' secretaries who become their nurses and *alter egos*; however, it is usually present in the way all women work, since it is difficult for women to change the 'bad' habit of nurturing. Without undermining solidarity as a positive value, the large majority of women we have met are highly sceptical about the value of this attitude, since even when formally recognized (as per the secretaries), it leaves the women in a subordinate position. However, they recognize that this skill is sought-after and could entail some economic advantages.

Finally, let us address two characteristics typical of the Commitment Universe, a broad discussion which involves on the one hand some of the present trends of company structure evolution, and on the other, the relationships between women, company structure and power. We characterize the first through the dichotomy intuition versus rationality, and the second through the dichotomy cooperation versus competiveness. With regard to the first dichotomy, let us quote Herbert and Stuart Dreyfuss (1986):

> . . . intuition is the product of deep situational involvement and recognition of similarity . . . Intuition or know-how, as we understand it, is neither wild guessing nor supernatural inspiration, but the sort of ability we all use all the time as we go about our everyday tasks, an ability that our tradition has acknowledged only in women, usually in interpersonal situations, and has adjudged inferior to masculine rationality.

Before them, Simon stated that intuition is the main mechanism which the manager has for overcoming the bounded rationality with which he has to deal. These two claims do not suggest that women can become managers without adopting 'rationality tools' like mathematics, computer science or economy foundations (Simon, 1977). Nevertheless, they argue that intuition is an important '*atout*' women can play.

The dichotomy cooperation versus competition, or better still, a good balance between the two, is a crucial point for the companies in the 1990s which will choose the network structure. In fact, they have to be both highly competitive in the marketplace, and highly cooperative and synergic with regard to the mutual relationships between the network companies. The dimension of the latter will possibly decrease and large hierarchical structures will be broken into smaller and more effective units. The increasing attention devoted to the development of the aforementioned Information Technology (IT) supporting cooperative work, confirms this trend. Such organizational and technological contexts could make more likely the basing of the individual career on acknowledgment of increasing competence, greater autonomy and relational abilities (cooperation), without necessarily surpassing other colleagues (competition). On the one hand, the described evolution could reduce the discomfort a number of women feel with respect to the male-oriented organizations and work culture of today (De Cindio and Simone, 1985), but on the other it would oblige them to deal with a less 'protective' organizational structure (non-hierarchical) and to be personally involved in the company power structure.

All these considerations allow women to see their characteristics as points of strength. In particular, they fully explain why women do not achieve careers. In current organizations career is tied to responsibility, and responsibility in its turn is defined in terms of the number of people one has under his or her control rather than in terms of his or her competence in fields of activity.

It also explains why the organizational and technological change we can foresee in the near future gives women an opportunity to change this unpleasant situation. Professionals are more likely to become key figures than middle managers, and companies will therefore be forced to shift the formal system of skill evaluation towards a greater acknowledgment of professional competence. Furthermore, the trend is toward the automation of repetitive (elementary) tasks, while people will be assigned the 'richer' tasks of handling break-downs in the context of the various socio-technical systems to which people belong.

The Universes in the Action/Research

In this section we discuss in more detail our experience in the field of affirmative action, which because of its richness, continuity and cooperation between research and intervention, offers the most complete setting in which to situate our thoughts and experiences. The following considerations and examples are mainly derived from the experience described in Osimo (1988), Osimo and Simone (1989) and Azioni Positive all'ITALTEL (1988, 1989, 1991).

The formative stages previously discussed are fundamental to the establishment of a correct and advantageous relationship between the women who form the focus of the intervention and the people charged with the intervention. Not only do these events allow the participants and 'secretaries' to get to know each other, more importantly, they allow the development of a relationship of mutual trust essentially based on two factors. First, there is the unequivocal demonstration of the different roles, and the varying 'expertise' which the women and the researchers bring to the exercise. The women are experts in their work, in their work environment and in the various workings of their organization, expertise which only those with long personal experience can have. The researchers bring expertise in analytical methodologies, are able to propose interpretive models of the phenomena discussed, and are able to offer alternatives to work situations which are found wanting.

Second, there is the clarification for all involved of the interpretive models used during the exercise, so that these become a point of cooperation and allow all to know both the stage and the subject of the analysis at any given time. In other words, each question and each answer are naturally placed within the overall schema, thus reducing the chance of attention wandering into pointless digressions, and of information contained in scattered or confused accounts being lost. This process greatly enhances the eliciting and acquisition of information.

The relationship of trust between participants is essential when the analysis involves small groups, where the technique of collecting information, be it one-to-one or in a group setting, involves keeping the participants constantly involved. In this setting the interpretive model based on the Universes has several distinct advantages: it is sufficiently powerful to deal with all aspects of work (from the use of technology to organizational factors, from issues of work layout to subjective factors). It is modular so that each phenomenon discussed can be considered both within the various dimensions, and in the totality of the analysis; and it is easily understood, thus allowing all the participants to be active and creative contributors to the analytic process. This last aspect is the main reason for our emphasis on the mutual trust among all people involved in the interventions. Perhaps this emphasis is strongly dependent on the fact that the interventions were mainly conducted by women. In our experience, male consultants/teachers very much enjoy playing the role of 'quick and omniscient problem solvers', whose main duty is to provide 'absolute truths' and 'universal interpretations of any phenomenon', and consequently to identify 'the best solution for every problem'. Recognizing that 'complementary competences' are needed to overcome limited knowledge is likely to be a 'female' characteristic acquired in the daily exercise of making compatible the requirements of different subjects (within their different lived social settings), denying neither their intrinsic individuality, nor introducing a fictitious hierarchy of relevance among them.

In the following sections we will consider how the model has contributed to the analytic phases which follow the initial training. We will do so by examining the three essential participants of any affirmative action intervention: the workers, the management, and the people undertaking the intervention.

It is to be noted that only the first group has specific training in the techniques used; the other two groups used the techniques implicitly, to the extent that we all had them as 'cultural baggage', which influenced our intervention and cooperation. We emphasize that what follows is based on our experience and is informed by our perspective as information technology experts.

The universes and the workers

The opportunity and ability to disassemble and reassemble the components of one's work allow a worker to examine that work from perspectives not previously possible. Certain aspects of the work which were previously thought important become less obviously so, while other aspects previously thought insignificant become of considerable interest. Some examples may help clarify this point, and give a rough idea of how our teaching methodology operates in the field.

Among clerical workers (who have formed the majority of our subjects), there is a natural tendency to think of their work as entirely routine, with negative connotations. Thus it is natural and appropriate to start the analysis along the Procedural Dimension; this allows one to show that the work has other dimensions, which may carry positive values, and to evaluate their mutual relationships. The choice of the Procedural Dimension as the starting point of the analysis is also due to the fact that it is in this dimension that (on paper) the majority of organizations, and information systems used in them, are designed. So, when it is necessary to develop an analysis of the organization and identify, along with the women, possible changes and reorganizations to be made, then time and effort must be put into developing a more detailed analysis of the work procedures. With this aim in mind, the teaching approach we have used for some time now, broadly described in De Cindio, De Michelis, Pomello and Simone (1983), proposes an office procedures representation language, which is quite easy to read, and can be used by the workers themselves. At the same time, it is powerful enough to represent the necessary knowledge (exchange of information among distributed executors, by whom and when the decisions are made, and with which consequences on the workflow, assignments of resources to the executors, and so on) with an intuitive and unambiguous interpretation (being based on a formal theory which does not however, appear to the user). Trying to 'formally' describe the office work flow: a) allows the workers to enrich individual knowledge of the procedure based on personal experience, with other employees' know-how; b) often allows a very 'alive' debate based upon a shared view of the work procedures, and c) provides the workers with a new language which it is possible to use in communications with the IT specialists in charge of the technological/organizational changes.

It is important to notice that during the analysis of the existing situation and its redesign, it is the 'secretary's task to ensure for the audience when and where the Procedural Universe is to be left, usually for the Operational or the Commitment Universes. This happens when the phenomena under consideration cannot adequately be represented by the previous descriptions which highlight where in the procedures exceptions can arise (Commitment Universe), or equipments which are inadequate (Physical/Operative Universe). For example, one can compare the importance of an inadequate interface (like the famous 'screen windows' typical of the Operational Universe) with the design of the task using this interface (Procedural Universe), thus dealing with unease about, and fear of, the use of computer technology in the appropriate domain. Further, one can examine the respective merits of having few terminals or personal computers (perhaps poorly sited ergonomically and with predetermined and inflexible access times), or of having a terminal or personal computer on each desk, to be used in conjunction with other activities and at flexible times determined by the user, perhaps with minor changes to allow a personalization of each worker's own space (Operative Universe).

Another example concerns sources of interruption of the normal work flow, due to the lack of information necessary to complete a task. This exceptional situation engenders the need for additional communication to acquire the missing knowledge. So, it is necessary to leave the Procedural Dimension in favour of the Commitment Dimension. Here a relevant question is: from whom can such information be retrieved? Answering this question highlights the actual information flow, which sometimes contrasts the official definition: the role played by the different organization members (specifically the management versus the experienced workers), with the need for suitable (technological) supports for preventing, whenever possible, these unwanted situations.

These insights are not only important in themselves as instances of personal growth, but are also helpful to workers defining a strategy with which to defend and augment the value of their work in the context of changes prompted by technical, organizational or market-place factors. An important example of this involves an understanding of the effects of technological change; what do these changes incorporate, what do they leave to the worker, and what is now provided or required which was not before? The use of the Universes allow one to consider all the dimensions of the change, and consequently to correctly address and answer these questions, in which case, the technological changes can be confronted with an attitude which is neither passive resistance, nor opposition aimed at defending the *status quo*, but instead as stated above, defines a strategy by which to plan the forthcoming battles.

The Universes help to understand in detail the gains and losses entailed in the new developments, and therefore where to focus not only one's defences, but more importantly requests for new equipment or the development of one's duties. For example, it is futile to entrench oneself in the defence of skills such as: manual precision, the capacity to remember information, or the ability to perform boring and repetitive tasks, since these are all better done by machines. It is far better to know where the need for human control (with its unmatched capacity for memory), attention, creativity and foresight, can make the work process more reliable and efficient; and to defend these factors from a knowledge of the ways in which they mesh with automation.

However, there are still some reorganizations which require a change of work duties, and in consequence of the tools used, these changes always bear a great cost. At times, however, it is still better to accept the cost, if the changes also lead to an increase of skills and interrelations with other parts of the organization. Here also, an understanding of all the dimensions of one's duties allows a judgment as to how proposed changes may alter them, and thus leads to the ability to request the retention of the tools and work methods which are still of use, and to request more appropriate ones where necessary.

The universes and the management

During the phases of the action/research described above, contact is frequently made with representatives of management, who because of their training, experience and the culture of management, often have very strong attitudes to and beliefs about the nature of work. These views are often stereotypes. Clerical work is primarily seen as the routine carrying out of decisions made by management, subject to occasional monitoring and control by the management hierarchy. Analogously, management work is seen as unstructured decision-making and control. Even if management accepts in principle new perspectives on the nature of work, the more traditional attitudes emerge strongly whenever a specific instance is analyzed.

This culture of management, with its inherent stereotyping, greatly reduces the possibility of accurately examining phenomena, and leads instead to a tendency to reject alternative values and perspectives based on the study of workers, deeming them 'subjective'. In other words, managers also like to play the role of 'quick and omniscient problem solvers' and enter the situation with predefined interpretations and solutions. They instinctively deny that direct interaction with the workers results in a view of the organization which is as worthy of consideration as the view of management. A classic reaction is: 'Your approach takes too much time and, moreover, information from the field is biased towards the workers' standpoint' — especially if women are involved in affirmative action programs — 'so this work is a waste and not usable'. The situation created is frequently onc of a clash of differing opinions, of deductions derived from different 'axioms' and by different 'inference rules'.

An example of this type of situation is as follows. For some time the various levels of management had not accepted that a group of employees (as part of their work) should spend a significant amount of time on the telephone speaking to people from outside the company trying to resolve exceptional cases. The reasons for this objection were two-fold: first, in their view, there were no 'exceptional' cases, and second the duties of the employees did not include such activities. When confronted by findings which emerged from enquiries and interviews conducted both in the workplace and with employees, management accepted the reality of the situation. However, perceiving the problems created by the situation as a 'mere interruption of the work-flow', they proposed a solution aiming at the elimination of the exceptions which triggered the problems. After long discussions, where we were able to prove that the use of the telephone could help in solving exceptions which would otherwise be insoluble, they agreed that it was more realistic and efficient to support the workers in this activity, and decided to offer them both technical and people support.

In such a situation of conflict, it is useful for the consultant involved to have access to an analytical and interpretative reference model with the characteristics of the Universes. The use of the Universes leads to a more balanced discussion, that is, a comparison of different theories rather than a confrontation between theory and deductions based on partial and subjective analyses. In addition, their use allows the discussion to occur on different levels ranging from the technological to the organizational to the subjective, avoiding the pitfalls of confusing these different levels, and losing sight of the fact that each is of equal importance, containing as it does, both part of the problem and part of the solution.

This matter of conflicting cultures is even more evident if one looks at interventions in the field of affirmative action where both the group being studied and those carrying out the study are mainly female, while the management is invariably predominantly male. Even in cases of affirmative action this is a 'handicap race'.

The universes and the action/research team
A good action/research team contains people of diverse professional backgrounds and work experiences. Each member should bring to bear a different perspective and should make this perspective known during the phases of enquiry and analysis. At times this will lead to a lack of understanding, to differences of opinion, and above all, to the risk of losing information and insights essential to an understanding of both the phenomenon under examination, and the possible solutions to the problems which arise.

Reflection upon our own participation in action/research teams has underlined for us the benefits of having a model such as the Universes through which to gain the cooperation of people with differing experiences from ours. The independence and complementarity of the work dimensions upon which they are based has encouraged us to perceive the contributions of others, not as 'disturbances' to our own elaborations, but as contributions, which if not always immediately accessible for integration, are surely positive, in pursuit of a collective interpretation process. Here we are not referring to a tolerance or politeness in our dealings with other professionals, but instead to a precise methodological attitude, based on a systematic decomposition and recomposition of the phenomena under consideration, along the dimensions of the Universes.

The final product is not only a non-conflictual attitude, but also an effective method of collaboration which allows (even in the most difficult moments of the enquiry), a precise method of integrating the various perspectives into a comprehensive whole, both during the preparation and the carrying out of interventions.

Conclusions

In this chapter, we have proposed the Universes of Discourse as a tool for analyzing the various dimensions of work and its computerization. We have described their application for developing a framework which identifies and discusses, via specific case studies, the work attitudes and practices of women. Finally we illustrate the ways in which this framework has been used in teaching and action/research programs devoted to women.

In relation to the latter part of the paper (specifically, the last main section), an attentive reader might ask (as we have asked ourselves), what is specific to women in our presentation? We have spoken primarily of work, of the organization of work and of technological changes. We argue that what is specific to women is that in a number of offices (and in some professional areas), it is women who predominate, because of the historical stratification of women's work.

Our experience has demonstrated that working with a group of women, or analyzing the work of women, and therefore to some extent being constrained to take their standpoint as the 'entry point' for the intervention, has led us to obtain information which the use of other standpoints would either not have revealed, or revealed only at the cost of greater effort and time. Although frequently involved in other organizational redesign processes connected with computerization, we have never obtained a comparable picture of the organization as it really is, of its 'troubles' and of its 'needs', which compares with that obtained after taking women's work as the entry perspective for the analysis and design.

We do not have a 'scientific' explanation of this empirical observation, but suspect that the reason is that women as either analysts, or the subject of analysis, are in general more favourably disposed to an understanding of the 'true' functionings of organizations. The characteristics specific to female work we have discussed above concerning the Procedural and Commitment Universes, for example, the concept of intuition, their attitudes towards cooperative behaviour and respect for other individuals' competences, and finally the attention to interpersonal communication, might indeed be viewed as contributions towards a better formulation of this intuition. Indeed, we believe it to be extremely challenging that the most significant and even less well recognized women's characteristics are identified at the commitment level. With some optimism we can say that the future offers women big opportunities for becoming the designers and managers of the new organizations, and of the new technology which supports them. It will be mainly their choice to enter or not in this game.

This intuition leads us to conclude this chapter with two provocative questions. First, that the difficulty in achieving success in affirmative action programs is due perhaps to the fact that the involved social groups are inclined to pay attention to 'classically female but negative' factors concerning women's work. Examples include 'maternity', lack of flexibility in the work time scheduling, reduced availability for travelling, female absenteeism etc. The same groups refuse to consider broader factors such as the organization of work, the content of work and the tools used. Could women teach something different to the traditional 'masculine' style of company management?

Therefore, the second question/claim is: do you wish to study organizations as they really exist? Then examine them from a women's perspective. Do you really wish to improve these organizations? Then again you should do so from a women's perspective.

Acknowledgments

We wish to thank G. De Michelis with whom we created the instrument presented; all the women whom we met, and with whom we worked in training groups and during research and intervention programs; the Affirmative Action Group of ITALTEL for their contribution to our understanding of women's issues; M. Sacerdoty for his help in the revision and translation of this paper and the editors for their useful and stimulating suggestions.

Note

1 Affirmative action programs for equal opportunities in Italy have recently been regulated by law (n. 125, April 1991). The law formalizes a previous practice, mainly stimulated by the Italian TU which allowed for many experiences of research/action interventions, both in public administration and in private companies. These programs facilitate the joint efforts of management, Trade Unions and external consultants to develop concrete policies to realize equal opportunities in specific organizational settings.

References

Azioni Positive all'ITALTEL (1988/89/91) *Uni ricerca intervento sul lavoro femminile* (I, II e III rapporto), ITALTEL, Quaderni.

Bagnara, S. (1987) 'L'Ergonomia del Software: Una condizione per il successo nei processi di automazione', in Bagnara, S. and Stajano, A. (Eds) *Software Amichevole*, Milano, CLUP.

Bikson, T.K., Halasz, F. (Eds) (1990) Proceedings CSCW90, Los Angeles, CA, ACM Inc.

Butera, F. (1987) *Dalle occupazioni industriali alle nouve professioni*, Milano, Franco Angeli.

Catasta, A., De Cindio, F. and Simone, C. (1986) *An Educational Experience toward Actions for Equal Opportunities*, presented at the 2nd Conference, Women, Work and Computerization, Dublin.

De Cindio, F. (1986) *E il computer parlera' al femminile*, il Sole 24 ORE, 122.35.

De Cindio, F., De Michelis, G. and Simone, C. (1987) 'Dimensioni dell'Usabilita', Universi del Discorso e Software d'Ufficio', in Bagnara, S. and Stajano, A. (Eds) *Software Amichevole* Milano, CLUP.

De Cindio, F., De Michelis, G., Pomello, L. and Simone, C. (1983) 'Conditions and Tolls for an Effective Negotiation during the Organization/Information System Design Process', in Briefs, U., Ciborra, C. and Schneider, L. (Eds) *System Design for, with and by the Users*, Amsterdam, North-Holland.

De Cindio, F. and Simone, C. (1985) 'Women and work in the age of computers: (Many problems), one opportunity and two challenges', in Olerup, A., Schneider, L. and Monod, E. (Eds) *Women, Work and Computerisation: Opportunities and Disadvantages*, Amsterdam, North Holland.

De Cindio, F. and Simone, C. (1988) 'A Framework for Understanding (Women) Work and its Computerization', in Tijdens, K., Jennings, M., Wagner, I. and Weggelaar, M. (Eds) *Women, Work and Computerization: Forming New Alliances*, Amsterdam, North-Holland.

De Cindio, F., Simone, C., Vassalo, R. and Zanboni, A. (1988) 'CHAOS: A Knowledge-Based System for Conversing inside Offices', in Lamersdort, W. (Ed.) Proceedings IFIP TC8/WG8.4, *International Workshop on Office Knowledge: Representation, Management and Utilization*, Amsterdam, North-Holland.

De Michelis, G. (1987) 'Gli Universi del Discorso e l'educazione linguistica', unpublished note, Lombardia, IRRSAE.

Dreyfuss, H. and Dreyfuss, S. (1986) 'Mind over Machine', The Free Press.

Ellis, C., Gibbs, S. and Rein, G.L. (1991) 'Groupware: Some issues and Experience' *Communication ACM*, Vol. 34–1.

Greif, I. (Ed.) (1986) Proceedings CSCW86, Austin Texas, ACM Inc.

Greif, I. and Ellis, C. (Eds) 1987 *ACM Transactions on Information Systems*, Special Issue on CSCW, Vol. 5.2.

Greif, I. and Suchman, L. (Eds) (1988) Proceedings CSCW88, Portland Oregon, ACM Inc.

Novara, F., Rozzi, R.R. and Sarcinelli, G. (1983) *Psicologia del lavoro*, Bologna, Il Mulino.

Olson, M.H. (Ed.) (1987) *Office Technology & People*, Special Issue on CSCW 3, 2.

Osimo, B. (1988) 'An experience of Affirmative Actions at ITALTEL', in Tijdens, K., Jennings, M., Wagner, I. and Weggelaar, M. (Eds) *Women, Work and Computerization: Forming New Alliances*, Amsterdam, North-Holland.

Osimo, B. and Simone, C. (1989) *Azioni Positive: Riflessioni su di una Esperienza in Corse*, WITEC Conference, Capri.

Simon, H.A. (1977) *The New Science of Management Decision*, Englewood Cliffs, NJ, Prentice-Hall.

Simone, C. (1987) 'A language to speak about parameters for office systems evaluation', in Proceedings IFIP Conference on Languages for Automation, Vienna.

Tijdens, K., Jennings, M., Wagner, I. and Weggelaar, M. (Eds) (1988) *Women, Work and Computerization: Forming New Alliances*, Amsterdam, North Holland.

Winograd, T. and Flores, C.F. (1986) *Understanding Computers and Cognition*, Norwood, Ablex.

Chapter 11

Trade Unions, IT and Equal Opportunities in Sweden

Eva Avner

Introduction

This chapter opens with a brief overview of trends in women's office employment and trade union policies in relation to computerization. In the second part, an example of an innovative trade union project, designed to improve opportunities for women office workers, is discussed.

Trade Unions and Female Office Workers: A Brief General Survey

In Sweden white and blue-collar workers are organized in different trade unions. The unions have entered into agreements which stipulate which union is to organize particular groups of workers. These agreements encompass both the public and private sectors, and manufacturing. Thus, female office workers are found in different unions, depending upon the position of the organization in the labour market and the nature of the industry.

SIF[1] organizes all private sector office staff in industry and is the biggest white collar workers' union. There are close to 200,000 male members, mainly technicians, and more than 100,000 female members, most of whom are in office jobs. SIF is a vertical union with members from all categories of staff. Approximately 60,000 members are managers, mostly male.

TCO is an inter-union organization for certain areas of cooperation. All organized white-collar workers are affiliated to it through their unions. TCO have thus 1.1 million members of which 400,000 are women in a variety of office jobs, including 150,000 who work as secretaries. Other jobs include cash clerks, accountancy clerks, wage administrators, general office staff (mainly in smaller companies) and order clerks. About 80 per cent of the staff workers of Sweden are organized.

The national figures for women's employment are close to those for men: 84 per cent of the female population and nearly 90 per cent of the male population. Of these more than 30 per cent of the women work part-time (20–34 hours) while only 5 per cent of the men do. The higher numbers of women in part-time work can largely be explained in relation to the fact that women carry the main responsibility for child care and domestic work in addition to their paid work.

In Sweden parents can share maternity leave (eighteen months in total with 90 per cent salary directly after birth; this drops to a reduced amount in the last six months. Parents are also entitled to stay home when the child is ill with full pay for up to sixty days per year). This is reinforced by a special state parental insurance. Current statistics demonstrate that most children spend the first year mainly with their mother, (92 per cent of the time with the mother and 8 per cent with the father). Later on in the child's life the fathers tend to become more active parents and at present some 41 per cent of men take leave when a child is ill. (Figures taken from 'Women and Men in Nordic Countries: Facts on Equal Opportunities Yesterday, Today and Tomorrow', published by the Nordic Council of Ministers, Copenhagen, 1991.)

Currently, some 5 per cent of working women (over 100,000) are single parents with a child under 18 years; the corresponding figure for men is 0.6 per cent (15,000).

At present, women office workers are a priority group for many unions. This is because the office stuation is changing rapidly; new technology and new economic strategies have altered work organization and specific jobs. These changes are supported by new and more advanced computer systems that either make jobs more automated or move them from the woman who works as a secretary to the manager or to the professional. In Sweden, unemployment has increased more among women clerical workers than in many other occupations: in 1991, for instance, it was over 3 per cent while total unemployment was 2 per cent.

Women's unpaid work develops skills and experiences which they take with them into their paid work. They blend this knowledge in with their professional skills but neither work organization nor the design of new technologies seem to reflect these womanly qualities.

Important personal qualities for a secretary are: diplomacy, forward planning and 'juggling' a number of jobs simultaneously. Important qualities for a wage administrator include maintaining good relations with both fellow workers and managers, being meticulous and getting to know staff. Cashiers and bookkeepers also need to be meticulous, have a good knowledge of the company, and be able to judge the plausibility of the financial amounts recorded. All three groups need to be flexible and able to work well under pressure. At a formal level secretaries stress the importance of language skills to the job; wage administrators stress their knowledge of finance laws and wage agreements, and bookkeepers their knowledge in bookkeeping. However, aspects also essential to their work include their tacit knowledge and intuitive skills, which are not easy to put in words or enter into the computer (Göranzon, 1982). Important aspects of the work and skills which women in offices use and develop in their work remain unseen. Only when these aspects of the work are not completed do they become visible.

Women are socialized into taking major responsibility for the family, which includes developing the capacity to empathize closely with other people. Taking responsibility is a skill which women take with them into paid employment, a quality which is directed to colleagues and end products. Responsibility in relation to products and services is often translated into a desire for producing high quality work. 'Women's double workload demands qualities like good planning and a high degree of flexibility, qualities which most women contribute to their paid work' (Gunnarsson, 1989).

Specific 'womanly' skills which women as a group are able to contribute to paid work include high degrees of: empathy, assuming responsibility, forward planning, flexibility and close evaluation of the needs of others. These may be summarized as follows:

Socialization	*House Work*	*The 'Double Shift'*
— responsibility	— planning	— planning
— empathy	— flexibility	— flexibility
	— evaluation of needs	

(Andreasen and Jorgensen, 1987).

Currently, major efforts are being made to assist women to find acknowledged methods which help them gain greater recognition in their paid work. A key example involves the use of trade union study material, which supports women in finding methods through which to analyze their working situation. Being able to reflect upon their lives, their own skills and abilities, enables women to understand the barriers, to work towards overcoming them and to plan for a more 'empowered' future. Parallel research and development work on gender and work evaluation is being progressed in trade union forums (Acker and Ask, 1989).

The Office of the Future, A Swedish View of the Organization of Work

It is often assumed that new economic strategies lead to more efficient decision-making and a more flexible organization of work. One competes with time and quality at the same time as one wants to reduce all indirect costs. It means that stock is a 'forbidden' word, that all decisions are made on the right level, that is, each decision is closely connected to its execution. Indirect work is regarded as a burden, 'just in time' is a word of honour, broadly skilled people become attractive. What does such a work organization look like? How does it affect office women?

Studies of women in the Nordic countries include major debates about the gender elements of job sharing in a society where some theorists argue that women and men develop different rationalities. In the book quoted above, Ewa Gunnarsson highlights a model originally developed in unpublished research by Hildur Ve, as providing an explanation of women's limited access to leading the work of social change together with an explanation of women's subordination at work. Gunnarsson argues that this is related to the fact that the 'limited' rationality of technological progress assumes superiority over the rationality of responsibility in our society. A consequence of this, and of the portrayal of technical rationality as 'male' is that women will be more influenced by patriarchal ideas, rules, definitions and meanings than men will be influenced by 'womanist' ideas.

As we can see from the model below, this discussion was originally limited to reproduction and that part of working life where the work of caring is traditionally found. The model can be applied to the position of women in traditional offices, and it also begins to explain their new role in the new working routines which are on the way. Women in the office have always had the function of service: that of 'office wife'. In order to manage the job they have had to listen to and empathize with the needs of others. Most women have had to plan their tasks within a context of limited time and limited resources, alongside considerations for major responsibilities which they hold outside paid work. These dual responsibilities have encouraged the evolution of skills which are transferable to the workplace, such as forward planning and job flexibility; these qualities are being heralded as key attributes of the ideal worker for the 1990s. Therefore they have developed many skills regarding planning and flexibility related to the work responsibility and appraisal of needs. Flexibility, including the ability to assess the whole situation, and put the customer at the centre, are key criteria for the 'new' organization of paid work.

The model discussed by Gunnarsson is interesting to study in this perspective. It appears in 'Kvinnors Arbetsrationaliteter och Kvalifikationer' (Gunnarsson, 1989).

Gender-related work roles in society are the basis on which women and men develop different rationalities.

Industrial production of goods	Reproduction — the paid and unpaid work of caring
Limited technical rationality	Rationality of responsibility
The technical rationality has as its main aim efficiency, meaning that people and machines together produce as much as possible using as little energy and time as possible.	The rationality of responsibility has as its main aim to create a state of being, physically and mentally for those in need of care. The aim is also often to develop an ability to manage this aim.
Economic rationality has as its main aim to increase the accumulation of wealth.	
Bureaucratic rationality has as its aim an efficient administration fulfilling the aims of the technical and economic rationalities.	

The new economic strategies aim to increase the accumulation of capital, and this presupposes that a minimum of time and energy is consumed. So far the bureaucratic rationality has been harder to integrate into the rest. The office has been regarded as the bottleneck. Advanced computer systems for the administration are developed faster and faster. The work organization becomes levelled, hierarchies become obsolete, the staff becomes more autonomous.

By means of different work organization and more sophisticated computer systems, more of the work can be done at the source, that is, decision and action are brought together by means of integrated computer systems, which also enable management to maintain an overall picture by means of the computer system.

A system of work organization, based upon the rationality of responsibility, puts the well-being of people first, whereas a system based on limited technical rationality equates 'man' and machine, and puts 'womanly' aspects of organization in opposition to the system. The female ways become opposed to the male.

Flexibility, seeing the whole picture and putting the customer at the centre, is the basis of work organization underpinned by theories of capital rationalization (new economic strategies). Women develop high levels of flexibility and the ability to survive under pressure. They are accustomed to thinking holistically (meshing employment and domestic arenas), seeing the whole and able to take in the surroundings which means that women workers will be attractive to employers in future job markets.

However, women's demand for 'one-ness' and concern with care also takes in broader aspects of paid work, such as, life in the organization, workmates, etc. Available experience shows that in group-oriented work, where time and energy are minimized, most women are at a disadvantage. They prefer to consider work tasks in relation to domestic and caring responsibilities, (for example, allowing for being home with a child who is ill) and may therefore upset the planning of groups which have not planned for unforseen events like absence, because their organization is based on technical rationality and not on the rationality of responsibility.

The project described below has the aim of women learning to see and understand their own work situation, in order to be able to influence paid work and its development in partnership with the local trade union. First, we need to consider at a general level the ways in which office information technologies and working patterns are changing.

Technology: Background Issues

The use of computers has increased heavily in the 1980s, at the same time as changes in why and how computers are used at work. Within the next decade most of the paid workers in Sweden will work with personal computers. Many companies have systems already which are increasingly networked with the outside world and used as multi-user systems through local networks.

It is predicted that most professional groups will in the future work with personal computers; a knowledge of computers will constitute an important part of their professional knowledge.

Existing levels of computerization have led to the integration of many functions, especially on the shop floor, but this is also occurring in the sphere of administration. Office functions are built into different computer systems, and tasks which used to be carried out by different clerical groups are increasingly carried out by managers and professionals.

Many different professions are developing new roles. For example, foremen and intermediate managers may lose their roles as work organizers and supervisors; secretaries may be given less typing work; the wage clerks may not calculate the wages of employees for much longer, and coding and accounting will no longer be the sole province of accountants. Managers and varying professionals already type their own manuscripts into their word processors. They feed their absences into the wage system, complete their own travel accounts on the computer, and enter invoices into the computer after they have been checked by means of the computerized budget system.

The border-line between managers and professionals on the one hand and staff on the other becomes more diffuse. Will this be advantageous to women?

For the first time historically there is reason for hope. The fear of lack of skilled labour due to the diminishing numbers of children born, and the fact that the need is for people who are creative, who consider the whole, who are flexible and multiskilled is encouraging. Women's special skills fill these demands. If women through their experiences can contribute towards a work organization, which also includes rationality of responsibility, then perhaps we are heading for an exciting work life.

The increased use of computers in expanded areas of employment carries with it a prerequisite of user-friendliness if the systems are to prove efficient. New types of programming languages are being developed, making it possible for the user to do some of the programming. The interface between person and machine is a high priority research area, linked more to knowledge of how humans function and act in particular surroundings. The computerization of the office affects women in a variety of office professions and suggests the following questions: What jobs will women be doing in the future? Is it possible to create computer systems that make use of the special qualities that women have? Is this indeed a desirable ambition?

Strategies

The unions' strategies aim to help women become visible and active at the work place. One example is the project described below. The women's awareness of their strength and their knowledge is the driving force in the moves towards change, which will benefit both women and men.

SIF has since 1984 worked with the project JDP, where the letters (in Swedish) stand for Equality, Computerization and Personnel development. Every year one or two projects are undertaken involving four to seven union branches in each project. Our strategy is to develop methods, start projects and study circles and thereby to support women in gaining greater acknowledgment and becoming more involved in the process of change in the workplace.

A Successful Project: Background

The following are likely to be essential elements of the working life of tomorrow:

—a faster tempo;
—a higher degree of computerization;
—more abstract elements within jobs;
—parts of the office moved out to the workshop, the car, the home.

Most of SIF's female members have administrative jobs. The current office automation influences their job content and the organization of work, and this influence is likely to increase. For many it will mean that the job changes; for others, jobs will disappear altogether. For all these changes to lead to better working conditions, women themselves have to be involved in the process of changing the organization of work. However, this does not happen automatically. The majority of these women have long work experience and thereby a sound knowledge of their jobs. Still, they tend to underestimate their knowledge. Probably one of the reasons is the character of office work: women often perform tasks required by others, while others judge the importance of a certain task and set the deadline for it. The general relationship between office work and female socialization was discussed at the beginning of this chapter.

The JDP project was developed to stimulate and support women to get involved in the development of their own profession.

What We Wanted to Achieve by the Project and What We Hope It Will Lead to

Our aims:

- to find a model that would make unions more effective in their work relating to new technologies at the work place;
- to increase women's knowledge of their own potential, i.e. to get women involved in the development of their own professional role in order thereby to influence the development of the office of the future;
- to increase knowledge of this field among local union committees in order to enable them to participate from an earlier stage in the introduction of new technologies in the office.

We wanted the project to lead to changes in the organization of work and in office workers' tasks, to enable members to gain increased influence and expanded duties as computer use increases.

Our methods: What is 'project work?'
Berthelsen, Illeris and Poulsen, three Danish behavioral scientists on whose ideas we have based our work, write:

> We define project work here as a teaching method in which pupils, in collaboration with the teacher and perhaps others, explore and tackle a problem in close contact with the practical situation in which it is experienced. This means that the work will lead to ever wider perspectives and ever deepening insight . . . The role of the teacher is not merely to impart knowledge but primarily to act as an initiator, an inspiration, an establisher of frameworks and a consultant.
>
> The work is to result in a concrete product, which may be an oral or written report or may be formulated in other media or actions.

Project work is not a new invention. Most of us have been taking part for many years in various groups entitled 'project groups'. In fact, the underlying educational ideas can be traced back to debates in the mid-nineteenth century. (For some examples in the UK context, see Mace and Yarnit, 1987.) What has happened in recent years is that well-known educational concepts have been made more precise and perhaps more concrete.

Our method of work stands on the following four legs:

1 It is problem-oriented, as distinct from subject-oriented. The participants describe and deal with their problems as they see them in their daily situation at work.
2 It is participant-controlled, as distinct from leader-controlled. This means that the project leaders set up a form or framework for the work. The participants' control is on the level of content — the choice of areas and problems to be dealt with.
3 It demands the production of documentation, as distinct from merely consuming knowledge. This implies that all participants must deal with some aspect of the reality they experience at work. As an essential part of this, they must agree on a joint product — at least a written report. The demand for a product is important; it compels every participant to consider and understand a problem in depth and then stand for this analysis together with the others.
4 It focuses on collective processes, as distinct from training courses which focus on individual needs. The group's motive force and working material are to be its common interests. Individual needs and wants and private conflicts must therefore take second place to them.

For the participants in the project, the work offers practice and experience in:

- perceiving a connection between education and change;
- structuring an extensive body of material;
- organizing their own and others' work;
- cooperating in order to accomplish a task;
- taking responsibility for their actions;
- taking up a position and defending it;
- seeing objectives and perspectives in their work.

Participants

It has been easy to get the local union branches interested in the project. At this level there has been a perceived need to learn about and to become more active in connection with women's work situations and office automation. However, we often find it hard to recruit participants. After many contacts with regional offices and local unions within the federation we usually end up with three to five local union branches participating. The prerequisites for participation are

- that a new technology is being introduced at the office;
- that the local union has at least twenty-five women members working in clerical positions;
- that a functioning local union exists;
- that the local union is interested in the questions;
- that every participant sets aside around twelve working days, half of which will be used for seminars and the other half for 'homework' between the seminars;

— that at least three members of the local committee constitute one project group and between three and five women not holding union office form another project group.

How the projects operate
Participants learn by doing. The project goes on for a minimum of six months. The participants meet at three seminars to compare notes, to establish a theoretical basis and plan a pattern of activity. The work is carried out between the seminars. For the work to be successful the project is based on a form that provides support and is easy to take part in. The participants concentrate on the contents, receiving written instructions that help them to plan and structure the project work.

The two first seminars are used for theory in the form of lectures, ideas, discussions, role plays. Subjects include participation in decision-making, new technology, equality issues, and how to determine training needs. The subjects are chosen according to the types of problems the participants have declared that they want to work with. In the third and last seminar the different groups present their reports. All will have read the others' reports, which is a very important part. The reports are usually well-written, particularly the women's group reports; to ensure that they can be used at the place of work, they are all discussed seriously and often some part of the report will be revised.

What Are the Advantages with a Project Like This?

This is an 'easy' way to work. The participants can feel secure with it, thanks to clear instructions and a well-defined project organization, but they own the contents themselves. This is important, especially for women who often administer projects or type out reports for others rather than themselves. The participants' efforts should be focused on the contents rather than on the form of work.

My experience as the project leader is that women quite easily find their own way to start group work. They can identify and describe their problems at work, and they are able to plan group activities. Those who are union representatives (mostly men), often find it more difficult to get started. They find it hard to identify their own problems, that is, as elected representatives of the women. The women have their own experiences of office work, but most union officials do not feel they have this practical experience. Initially they tend to describe the womens' problems in a changing office, rather than their own problems as union representatives in a changing office.

At the outset it is important to identify a few important problems to work on throughout the project. It has been mentioned before that it is important that the project is successful for the participants. The selected problems are therefore tested by the participants against three criteria:

- *The subjective criteria*
  You must yourself feel involved and feel the importance of the problem being solved.
- *The objective criteria*
  The problem is to concern the whole professional group or department. Compare 'the four legs' of the collective interest.

- *The action criteria*
  The project group must feel that changing the situation is a realistic aim, that the problem can be dealt with.

This is perhaps the key step of the project. Some problems are left out if they do not comply with all three criteria. This is hard work, especially for the union representatives. When they discover that they have problems of their own, then the project is under way!

The project is rounded off with a seminar where each group presents a report on their project. We have found that the reports written by women are usually more personal and better written than those produced by the union representatives. To illustrate these points, here are some extracts from project reports, first from a women's project group, and second from a union branch.

Extract from a women's project group report
Conclusions
Information to female subordinates about new technology being introduced varies within the company, but generally speaking the information is poor if indeed there is any. Decision makers and investigators in the company don't think it is necessary to inform the female staff that will be affected by introduction of new technology and new equipment (or a proposed introduction). This is also true in the case of reorganization and introduction of new routines. We have concluded that women know that their possibilities to influence and benefit from reorganizations are nonexistent.

If the development continues in the same direction, then women in the future will have even less opportunities to influence their work situations, because they do not see the whole picture of what they work with, and computer people (mainly men) decide how the programs, systems and routines that women work with will be designed. If women are to be have greater influence, they must get training in computers and become programmers. (But will that mean a lower status for these jobs?)

A few women will be trained within the company to be able to get more challenging and stimulating jobs, when and if present jobs disappear or become computerized. This is partly because their former training does not fit in with the company's line of business, i.e. in the technical area, and partly because it is difficult to dispense with traditional jobs (budgeting, administration etc.).

The women's sense of being unequal will increase, if they are working with routine jobs at a terminal.

If the women are to have a chance, we must get training and also inform future generations of the necessity of training within nontraditional women's jobs and in the computer field.

It is probable that there will be more part-time jobs at terminals held by women and that means even less opportunity for influence. 'Well, you only work in the mornings!' or 'Well, you only work in the afternoons!' — when it comes to lack of information. In that way the small opportunities to influence how and where computerization is introduced in the company, will be even smaller.

Women in isolated areas will sit at home doing terminal work hooked up to a computer in Stockholm or something like that. That will mean working alone with no chance to talk to women in a similar situation, as they will be miles apart.

The TCO demand in the brochure 'Terminal Work in the Right Way', p. 12, stating that intensive routine work sessions at terminals should be two hours per person and day, that will be hard to get accepted, at least in practice.

SIF's demand in 'The Computer on SIF's Conditions', p. 10, stating max two hour intensive terminal work and then other tasks, and maximum two sessions per day, should have a better chance of getting accepted.

In 'Reading Computer Screens', Arbetarskyddsstyrelsen (Swedish Board for Worker Protection) directive 136, has no time specification, only in general that if there is eye tiredness or sight problems, the work should be arranged so that the worker gets relief in form of rest periods or other work with less strain on the eyes. How many women can up hold their right towards a boss who says: 'You'll get used to it!' when confronted with complaints of eye tiredness and/or sight problems? Conclusion: only a few.

Experiences

The situation is not good regarding finding alternative areas of work within the firm, other than the traditional ones, which women can move to if their tasks diminish or disappear with computerization or if the person concerned wants more qualified work.

There are at least some minor cases where women have been able to stick to their opinions and even been able to change bad routines or get changed work within a work group or change the group.

We have made contact with other women within the firm, something which we had not been able to do before. We could talk about their work, got their views on how the information works, on the working environment, the work situation and so on.

How we intend to use our report

We have talked with and interviewed many, but not everybody due to lack of time, about their work situations and what they expect in the future in Bacho [a Swedish firm]. They only know about their own thoughts, not realizing that there are ten or fifteen others with the same ideas. It would surely be good for everybody, especially our female work-mates, to read our report and see what we have been able to achieve.

We will make our report available to the SIF branch leadership in Bacho so that this can be part of a basis in formulating an action program for the project JDP (Equality, Computerization, Personnel development).

Extract from the branch leadership report. We have drawn these conclusions:

For new technology to be able to work in the company the following is needed:

- Computer familiarity for all, which does not mean that all must be able to program, but they should know what a computer is, what it can and cannot do,
- Knowledge of the company's range of products and individual products (systems),
- Knowledge of routines,
- Knowledge of the company.

This is how we intend to use the report to spread our knowledge.

The report will be a basis for the discussion with the company about personal development to prepare for future changes. It will be a basis for a concrete long-term program, negotiated with the company.

Technical and administrative routines will be specified so that personnel resources can be allocated when it is practical and when the market and the economy permit.

The report will also be shown to our members, so that we can influence and be part of the action programme, without a sense of being steam-rollered by the company or the union branch leadership.

Consequences for the participants and the work place

If you run your own project for six months and write your own report of course you change your views in some way. Most projects have dealt with questions concerning vocational training connected to the future of the company where the participants are employed. Some projects have focused on the work environment, both in the physical and the psychological area. To mention some results:

- —Three women from one of the first projects started their own projects in their region in order to get more women to start 'JDP' projects. They applied for money at the Swedish TUC which they succeeded in getting.
- —At one workplace the project ended in a new working environment. The participants at that workplace carried out an investigation, which showed how badly their workplaces were operating, and they managed to get some problems resolved.
- —At another workplace, they started with an enquiry about the psychological climate at the workplace and they managed to start a discussion about how to support and acknowledge one other.

But most projects have ended with suggestions for better vocational training. Many of the women have started to study. General consequences include:

- —Nearly all women feel that they have got increased self confidence;
- —The women become more active at the place of work. There are women's meetings with many participants where problems of the workplace are dealt with;
- —Women get called to information meetings and get more internal information;
- —Better terminal working places;
- —Improved work organization;
- —Staff appraisal;
- —Individual training schemes;
- —Personnel training:
  - languages,
  - computer training,
  - company knowledge,
  - product knowledge,
  - business economics;
- —Study circles and evening courses are started, both in trade union and professional matters;
- —More women become active as union officials and negotiators;
- —Better relations between trade union leadership and members.

Conclusions

In this type of project, where the participants have a great deal of influence, and where they are 'the owners' of the project, they learn that they can make a difference, that they can cooperate. Women and men develop a new view of each other. The local shop stewards become aware of women as a qualified group, which they commonly underestimate. The women realize that their abilities are needed, and that they can contribute in many areas. Nowadays, with increasing unemployment among women office workers, the emphasis is more and more on the problems they face. Projects such as those described here challenge conventional assumptions about women office workers' skills and attitudes, and help to open up opportunities for them to play an active part in processes of computerization and organizational change.

Note

1 SIF is the Svenska Industritjanstemannaförbundet: the Swedish Union of Clerical and Technical Employees in Industry. Postal address: S-105 32 Stockholm, Sweden.

References

ACKER, J. and ASK, A.-M. (1989) *Wage Differences Between Women and Men and the Structure Of Wage-Setting in Swedish Banks*, Swedish Centre for Working Life, Stockholm, Sweden.

ANDREASEN, K. and JORGENSEN, E. (1987) 'Vem Saatt Kvinnor ar Svaga?' (Who Said that Women are Weak?) *Kulturgeografiske Hafters Skriftserie*, No. 17, Copenhagen, Denmark.

BERTHELSEN, I., ILLERIS, K. and POULSEN, B. *Project Work*, Copenhagen, Denmark, Teachers Training College, undated.

GÖRANZON, B. (Ed.) (1982) *Job Design and Automation in Sweden: Skills and Computerization*, Swedish Centre for Working Life, Stockholm.

GUNNARSSON, E. (1989) *Kvinnors arbetsrationaliteter och kvalifikationer*, (Women's work rationalities and qualifications) Arbetslivscentrum, Stockholm, Sweden.

MACE, J. and YARNIT, M. (Eds) (1987) *Time off to Learn*, London, Methuen.

Notes on Contributors

Alison Adam is a Lecturer in Computation at UMIST (University of Manchester Institute of Science and Technology), specializing in artifical intelligence. She did a PhD on the historical sociology of scientific knowledge at Sheffield City Polytechnic. She also spent several years in industry as a software specialist and then worked as a researcher on one of the Alvey projects at Lancaster University before her present appointment.

Eva Avner is a full time official of SIF (the Swedish Union of Clerical and Technical Employees in Industry) in Stockholm. Her area of responsibility is personnel development, especially for administrative occupations (mainly women). Current projects include: 'At the service of human beings', career development of women as computing professionals — a Nordic Study; 'Teaching and learning' which aims to get more local influence over relevant vocational training and the 'Women's College', further education for women in office occupations.

Susanne Bødker is an Associate Professor at the Department of Computer Science, Aarhus University. Susanne is interested in systems development and in the processes of use of computer technology. Her theoretical interests are multidisciplinary and include work with HCI based on the socio-historical tradition of Soviet psychology. These interests are reflected in her book *Through the Interface* (Erlbaum, 1990). In the 1980s she took part in the UTOPIA and COOP projects, out of which comes an interest in experimental design methods. Currently she is working with the DEVISE and AT projects about which she has written many articles which focus on active strategies for designers and users working together. Her other professional interests include gender issues, tailorability, and understanding design as learning. She lives in a collective community in Aarhus with her son Jonas.

Margaret Bruce is a lecturer in the Manchester School of Management at UMIST, specializing in design management, innovation and marketing. After completing her PhD at Manchester Polytechnic she spent a number of years as a researcher and lecturer at the Open University prior to her present post. She has been Principal Investigator on a number of ESRC projects and is on the management committee of UMIST's PICT centre.

Fiorella de Cindio is Associate Professor of Special Programming Languages at the Computer Science Department of the University of Milan. The main focus of her work concerns languages and methods for the design and the implementation of concurrent distributed systems. Moreover she has studied methodologies for dealing with office work automation and is a member of the team developing a knowledge-based system supporting group work (CHAOS, Commitment Handling Active Office System). In this framework, she has devoted special attention to changes for women's work, and has been involved in a program for affirmative action undertaken by an Italian telecommunications company.

Ian Franklin is an occupational psychologist in the area of human factors for a large government agency (The Employment Service). Ian worked on the Human Centred Office Systems project at Sheffield Hallam University as a research assistant from September 1988 to September 1991. Ian's first degree is in psychology with an MSc in Information Technology where he specialized in human factors. In his current post Ian is endeavouring to apply the ideas of human-centredness to the development of large in-house systems.

Eileen Green is a Principal Lecturer in Sociology at Sheffield Hallam University where she is also Co-Director of the Centre for Organization, Management and IT (COMIT). She has taught, directed research projects and published in the area of Gender Studies since the 1970s and is currently working on research in the area of Women Managers, Organizational Cultures and IT.

Joan Greenbaum is a Professor of Computer Information Systems at the City University of New York (LaGuardia College). Joan's original field of study was economics where she set out to study the impact of technology on jobs and skill, particularly with reference to women's employment. Over the years her focus has been on developing programmes, teaching strategies and research methods to help computer system designers and office workers find ways to develop systems that support better working environments. Her first book, *In the Name of Efficiency* (Temple University Press, 1979) looked at the division of labour in the computer field. Her recent work includes a book about participatory design edited together with Morten Kyng, called *Design at Work* (Erlbaum, 1991). She has three sons and lives in Montclair, New Jersey, but has been a frequent guest at the Computer Science Department in Aarhus, Denmark.

As a local government officer, **Mike Hales** worked on employment and technology-related strategy issues at local government level, with an emphasis on 'bottom-up' approaches to policy and management. As a Senior Research Fellow he is now involved with the design politics of CSCW (Computer Supported Cooperative Work), their connections with the gender politics of knowledge and questions of post-engineering design. Current address: Centre for Business Research, University of Brighton; email — mh55@uk.ac.brighton.vms.

Flis Henwood is a Senior Lecturer in the Department of Innovation Studies, University of East London. She teaches on an interdisciplinary degree in New Technology and coordinates the Women and Technology Option both within that degree and within the Women's Studies Programme. Her main research interests are in the field of gender and technology and her doctorate was concerned with the identification of discourses of gender, work and equal opportunities in relation to technology.

Sonia Liff PhD, is currently a lecturer in the School of Industrial and Business Studies at Warwick University. Her current teaching areas include personnel management and gender relations at work. She has previously worked at Loughborough and Aston Universities. Her research interests include patterns of technological change, gender relations and technical change, and equal opportunities policies.

Fergus Murray works as a Research Fellow at the Manchester School of Management, UMIST on the ESRC Programme on Information and Communication Technologies (PICT). He has published widely in the area of the politics of IT management and systems development. He is presently researching the development of inter-organizational systems and networks in the European financial services industry.

Peter O'Hara is currently Business Development Director of OLM Systems Ltd. Prior to this, he was the Information Systems Project Manager in the London Borough of Camden's social services department where he was responsible over a five-year period for the management of a bespoke development for an integrated database. This involved developing the department's IS Strategy in the light of the changes arising from the NHS and Community Care Act and the Children Act as well as designing the IT infrastructure, procedures and standards. His publications include *Progress by Design: An IT Strategy for Social Services* (1989) and *Managing a Decentralised IT Strategy* (1989).

Jenny Owen joined the Human-Centred Office Systems Project at Sheffield Hallam University as a research student in 1985, following a number of years working in community development and equal opportunities in the voluntary sector. She then worked as a research associate there from 1989–91. She currently works as a research associate at the Manchester School of Management, UMIST on the ESRC Programme on Information and Communication Technologies (PICT). She has a particular research interest in patterns of gender relations and of technological change within public sector organizations.

Den Pain is a Senior Lecturer in the School of Computing and Management Sciences at Sheffield Hallam University. He teaches Information Systems Engineering and his research interests include Human-Centred Systems. Prior to working at the University he worked in the computer industry as a systems designer having started his career with a spell as a programmer.

Carla Simone is an Associate Professor at the Computer Science Department of the University of Milan. Her research areas are the theory of concurrency, and the representation and analysis of organizational and cognitive process within the language/action perspective. Within the area of design of computer supported cooperative work (CSCW) she is responsible for the project CHAOS (Commitment Handling Active Office System). Since the 1980s she has been involved in educational programmes aimed at women workers about the technological impacts on work, and she acts as a consultant for a permanent committee involved in the monitoring and promotion of woman's work in Italian industry.

Juliet Webster is a Senior Lecturer in the Innovation Studies Unit, University of East London. She has carried out research on the development of word processing and changes in secretarial work, on the computerization of manufacturing management, and on electronic trading. She is the author of *Office Automation: The Labour Process and Women's Work in Britain* (Harvester Wheatsheaf, 1990).

Index